D1281348

PRENTICE-HALL FOUNDATIONS OF PHILOSOPHY SERIES

Virgil Aldrich	Philosophy of Art
William Alston	Philosophy of Language
Stephen Barker	Philosophy of Mathematics
Roderick Chisholm	Theory of Knowledge
William Dray	Philosophy of History
Joel Feinberg	Social Philosophy
William Frankena	Ethics
Carl Hempel	Philosophy of Natural Science
John Hick	Philosophy of Religion
David Hull	Philosophy of Biological Science
Willard Van Orman Quine	Philosophy of Logic
Richard Rudner	Philosophy of Social Science
Wesley Salmon	Logic
Jerome Shaffer	Philosophy of Mind
Richard Taylor	Metaphysics

Elizabeth and Monroe Beardsley, editors

PHILOSOPHY OF BIOLOGICAL SCIENCE

David L. Hull
UNIVERSITY OF WISCONSIN-MILWAUKEE

PRENTICE-HALL, INC.
Englewood Cliffs, New Jersey

Library of Congress Cataloging in Publication Data

Hull, David L
Philosophy of biological science.

(Prentice-Hall foundations of philosophy series)
Bibliography: p.
1. Biology—Philosophy. I. Title.
QH331.H84 574'.01 73-12981
ISBN 0-13-663615-8
ISBN 0-13-663609-8 (pbk.)

To R. A. W.

10 9 8 7 6 5 4 3 2 1

PRENTICE-HALL INTERNATIONAL, INC., London
PRENTICE-HALL OF AUSTRALIA, PTY. LTD., Sydney
PRENTICE-HALL OF CANADA, LTD., Toronto
PRENTICE-HALL OF INDIA PRIVATE LIMITED, New Delhi
PRENTICE-HALL OF JAPAN, INC., Tokyo

FOUNDATIONS OF PHILOSOPHY

Many of the problems of philosophy are of such broad relevance to human concerns, and so complex in their ramifications, that they are, in one form or another, perennially present. Though in the course of time they yield in part to philosophical inquiry, they may need to be rethought by each age in the light of its broader scientific knowledge and deepened ethical and religious experience. Better solutions are found by more refined and rigorous methods. Thus, one who approaches the study of philosophy in the hope of understanding the best of what it affords will look for both fundamental issues and contemporary achievements.

Written by a group of distinguished philosophers, the Foundations of Philosophy Series aims to exhibit some of the main problems in the various fields of philosophy as they stand at the present stage of philosophical history.

While certain fields are likely to be represented in most introductory courses in philosophy, college classes differ widely in emphasis, in method of instruction, and in rate of progress. Every instructor needs freedom to change his course as his own philosophical interests, the size and makeup of his classes, and the needs of his students vary from year to year. The nineteen volumes in the Foundations of Philosophy Series—each complete in itself, but complementing the others—offer a new flexibility to the instructor, who can create his own textbook by combining several volumes as he wishes, and can choose different combinations at different times. Those volumes that are not used in an introductory course will be found valuable, along with other texts or collections of readings, for the more specialized upper-level courses.

Elizabeth Beardsley / *Monroe Beardsley*

CONTENTS

5

PREFACE

Because this book is intended to be an introductory text, I have made it as comprehensible and self-sufficient as possible. However, I cannot pretend that anyone will find it easy reading. The writing is too compressed for that. There is too much that must be said and too few pages to say it in. The major problem is that few readers will have a background in both biology and philosophy. In fact, many are liable to have a background in neither. Anyone who has ever taught a course in Philosophy of Biology knows how difficult it is to teach philosophy to biology students while at the same time teaching biology to philosophy students. The strategy which I have adopted in this book is to introduce the necessary biology in the early pages of the book and then to ease gradually into the more philosophical issues. I have also tried to define each technical term, whether biological or philosophical, when it is first introduced. Space limitations have prevented my repeating these definitions when the terms recur.

To the extent that this book is both comprehensible and informative, I owe a strong note of thanks to numerous philosophers and biologists who have read various versions of the manuscript. Among them I wish to express special thanks to Morton Beckner, Dorothy Grover, Helen Heise, Richard Levins, Ernst Mayr, R. E. Monro, Rosario Morales, Michael Ruse, Kenneth Schaffner, Leigh Van Valen, Mary Williams, William Wimsatt, and the students in my Philosophy of Biology courses. I also wish to thank Elizabeth and Monroe Beardsley both for their editorial suggestions and for the opportunity to write this volume for their series. Dawn Klemme, Dorothy Wilson, and Linda Erich typed and retyped the manuscript. The research for this book was done in part under National Science Foundation Grants GS–1971 and GS–3102 and a summer research grant from the University of Wisconsin–Milwaukee.

David L. Hull

Introduction

ONE PHILOSOPHY OF SCIENCE OR MANY?

Traditionally the formal sciences (mathematics and logic) are distinguished from the empirical sciences (for example, physics, biology, and psychology). In the Foundations of Philosophy Series, logic and mathematics are treated in three separate volumes by S. F. Barker, W. C. Salmon, and W. V. Quine. Empirical science in general, the social sciences, and history are investigated respectively in the volumes by C. G. Hempel, Richard Rudner, and William Dray. The overriding question that pervades these latter volumes is whether the traditional divisions of the empirical sciences into separate disciplines like geology, astronomy, and sociology reflect only differences in subject matter or result from basic differences in methodology. In short, is there a single philosophy of science that is equally applicable to all areas of natural science, or are there several philosophies of science, each appropriate in its own domain?

Hempel in his lucid volume sets out what has come to be the received doctrine on the philosophy of the natural sciences. He argues that a single philosophy is applicable equally to all areas of science. Scientific inquiry is characterized by the hypothetico-deductive model. According to Hempel, there is no logic of discovery in the sense of a set of procedures by which true scientific laws can be generated automatically from data. Scientists discover hypotheses in many different ways. Although some methods of discovery tend to be more productive than others, these methods are so varied and their success so erratic that little can be said in general about them that is very informative. One thing is certain, however. The truth of a scientific claim is independent of its source. No method of discovery can guarantee truth, and a scientific statement can be true regardless of how it was generated.

Hempel argues that the situation is very different with respect to the logic of justification. Although scientists go about checking their scientific hypotheses in a variety of ways, their procedures share certain essential features. It is the investigation of these features of the logic of justification that constitutes the true subject matter of the philosophy of science. First and foremost, according to the hypothetico-deductive model of science, a scientific hypothesis must be testable. Ideally the hypothesis to be tested is universal in form (for example, all gases expand when heated) so that in principle a single disconfirming instance can falsify it, while numerous and varied confirming instances gradually, though never completely, confirm it. In actual practice, the hypotheses tested are not always universal in form and occur embedded in scientific theories, not in isolation. Thus, both confirmation and disconfirmation tend to be complicated affairs.

Within this same view of science, the only genuinely scientific explanations are those that conform to the covering-law model. According to this model, explanation is inference. A state of affairs is explained by being inferred from a set of laws and initial conditions. These inferences may be either deductive or inductive. In a deductive inference, the conclusion follows necessarily from the premises. In an inductive or probabilistic inference, the conclusion follows from the premises with only a certain degree of probability. Hence there are two types of covering-law explanations. In deductive-nomological explanations, the inference is deductive. In statistical-probabilistic explanations, the inference is inductive or probabilistic. Thus, explanatory force is purely a matter of the degree to which the inference in the explanatory argument approaches the deductive ideal. The characteristics of the deductive-nomological model have been set out in some detail. As one might expect from the grab bag definition of "inductive" as any legitimate inference that is nondeductive, statistical-probabilistic explanations are a mixed lot. No single model has yet proved adequate to explicate any one variety of nondeductive explanation, let alone all of them in general.

THEORIES AND THEORY REDUCTION

Equally central to the logical empiricists' conception of science is the analysis of a scientific theory as a set of inferentially related statements. A few of these statements serve as basic axioms or postulates from which other statements, termed scientific laws, are derived. These laws in turn are related in very complex ways to phenomena that we as human beings can observe fairly directly. It is in this way that scientific theories are confirmed or disconfirmed. If in addition scientific theories are interpreted realistically (as they will be in this book), then some of the axioms will refer to certain entities, frequently unobservable, termed theoretical entities. In a realistic interpretation of scientific theories, these theoretical entities are thought of as existing in the same sense that objects of our ordinary experience exist, and scientific laws are viewed as reflecting actual regularities in nature. (For a defense of a realistic interpretation of scientific theories, see Simon, 1971.)

Traditionally, the subject matter of empirical science is grouped into areas of decreasing scope. Physics is thought of as having the broadest scope because it deals with the physical properties of all bodies and all bodies have physical properties. Chemistry is viewed as being only slightly less basic, because all material substances also have chemical properties; however, chemical properties are explained by reference to physical properties, and physics and chemistry become fused at the level of their most fundamental axioms. Biology is considered to have a more limited scope than physics and chemistry because it concerns only those physical objects which are also alive. All living organisms are physical objects, but not all physical objects are alive. Psychology is of even more limited scope, because it deals only with those living creatures capable of sensation. Sociology in turn is of even narrower scope, dealing only with sentient beings organized into societies.

Philosophers and scientists have used the term "reduction" in a variety of ways. Given the above analysis of scientific theories and the organization of the subject matter of science into the usual hierarchy, three senses of "reduction" can be distinguished with some clarity—epistemological reduction, physical reduction, and theoretical reduction. Epistemological reduction concerns the proper relation between scientific theories and the objects of our knowledge. Physical reduction concerns the relations between the fundamental entities postulated by various scientific theories. And theoretical reduction concerns the relations between the scientific theories themselves.

The goal of epistemological reduction is the elimination of any reference to theoretical entities in scientific theories. Instead, scientific theories are to be reformulated so that they refer only to the objects of our knowledge. There is some disagreement among epistemological reductionists over the nature of these objects. According to one version of epistemological reduction, all scientific statements are to be reformulated in terms of gross physical objects, usually measuring instruments like yardsticks and galvanometers. Another version specifies their reformulation in terms of sense data like "red patch now." The appeal of epistemological reduction stems from the empiricist claim that all empirical knowledge comes from sense experience; hence, it should be reducible to it. In point of fact, neither of these versions of epistemological reduction has met with much success. Nor do the issues raised by epistemological reduction have much to do with biology or vice versa. Accordingly, this sense of reduction will be all but ignored in what follows.

In physical reduction, systems at one level are analyzed into their component parts and the behavior of these higher-level systems are explained in terms of the properties, behaviors, and arrangements of these parts. The stock example of reduction to be found in the philosophical literature is the explanation of the gross properties of gases (like temperature) in terms of the movements of the molecules that make them up. Similarly, molecular geneticists are attempting to explain the behavior of genes in molecular terms. In theory reduction the axioms of one theory are derived as theorems from the axioms of another theory, and the derived theory is said to be reduced to the original theory. Again, the stock example of such a reduction is the derivation of classical thermodynamics from statistical mechanics by identifying the temperature of a gas with the mean translational kinetic energy of the molecules which make it up.

Given the preceding hierarchy of subject matters of science, the results of physical reduction and theory reduction tend to coincide. Both with respect to the scope of the relevant theories and the level of physical analysis, physics is basic. Physics deals with the physical properties of systems from the most highly organized beings to the simplest subatomic particles, whereas biology, for example, deals with the properties of only highly organized beings. Scientific theories are formulated at all such levels of analysis from the universe to evolving species to subatomic particles. A reduction is termed intralevel if both theories concerned refer to phenomena at the same level of analysis and belong to the same traditional area of science. If either of these conditions is not met, then the reduction is termed interlevel. Hence, the reduction of thermodynamics to statistical mechanics is intralevel in the sense that both theories are physical theories, but interlevel in the sense that the reducing theory concerns lower-level phenomena than the theory being reduced.

The preceding analysis of reduction has not gone unchallenged. G. G. Simpson (1964), for example, has argued that biology and not physics "stands at the center of all science" because "*all* known material processes and explanatory principles apply to organisms, while only a limited number of them apply to nonliving systems." The various branches of science are better organized "not through principles that apply to all phenomena but through phenomena to which all principles apply." Hence, if anything is going to be reduced to anything, it will be physics to biology. As Michael Simon (1971) has observed, if Simpson carries his line of argument to its logical conclusion, then the social sciences and not biology stand at the center of all science, because one could claim with equal justification that all known biological processes and explanatory principles apply to man, while only a limited number apply to nonhuman living systems. Hence, if anything is going to be reduced to anything, it will be physics and biology to the social sciences.

But the cogency of Simpson's argument and Simon's extension of it are marred somewhat by the fact that the relevant premises are false. It is simply not true that all known material processes in any significant sense apply to all physical systems, including organisms. For example, some physical bodies can emit electrical currents; others cannot. Some are magnetic; others are not. Apparently, some organisms can emit electrical currents, but none to my knowledge exhibit much in the way of magnetic effect. All Simpson can legitimately claim is that human beings do not contravene any genuine biological laws (where they apply), nor do living organisms provide any counter-instances to any genuine physical laws (where they apply). From these premises, neither Simpson's nor Simon's conclusion necessarily follows. To decide which science is central, a random sampling of natural phenomena must be taken to determine which science possesses more phenomena to which all material processes and explanatory principles apply. Simpson's mistake is that he does not go far enough in challenging the traditional ordering of natural phenomena. All he does is set it up on end. But this ordering itself is suspect, zeroing in as it does on man at its center. A magnet might produce quite a different ordering, centering no doubt on some special property of magnets. In spite of certain persistent doubts about the traditional anthropocentric organization of the sciences, it will be accepted for the purposes of the exposition in this volume. Not everything can be questioned at once.

REDUCTION AND THE BIOLOGICAL SCIENCES

Richard Rudner in his book on the social sciences argues for the extension of the received view on the nature of science sketched above to the social sciences. He points out that the social sciences differ from the natural sciences in their use of special techniques of discovery and even of justification, but employ the same logic of justification, and as far as philosophy of science is concerned, justification is all that matters. Rudner's exposition differs from that of Hempel chiefly in his use of the idea of partially and fully formalized theories to elucidate the nature of social theory. Unfortunately, because the social sciences have little in the way of explicitly formulated scientific theories, the technique in this instance is less informative than it has been in the physical sciences.

William Dray, on the other hand, argues against extending the Hempel-Rudner analysis of science to cover historical inquiry, because history is not a "science" in their sense of the word. The historian is concerned with establishing not only what human actions of societal significance have actually occurred but also with *understanding* them. The type of explanation and understanding provided by historians is different in kind from that found in other, more strictly scientific disciplines like physics. Similar opinions have been expressed by other philosophers and scientists about all the social sciences. Of course, the cogency of this position depends on an adequate analysis of these special senses of "explanation" and "understanding." I for one must admit that I have yet to comprehend these notions with any clarity.

The purpose of this volume will be to take a closer look at that area of science which has been passed over in the rapid extrapolation from physics to the social sciences. During the early years of science, certain modes of explanation, which are most appropriate to biological phenomena in general and human behavior in particular, were read into all of nature. Just as a man might strive to be virtuous and a species might strive to reproduce itself, a rock falling to the earth was interpreted as striving to attain its natural place. The opposite tack has become increasingly popular since the time of Galileo and Newton. A type of explanation that originated in the study of purely physical phenomena has been extended to biological and social phenomena. All events are explained in terms of antecedent events organized in causal chains and networks, characterizable in terms of universal laws which make no reference to the causal efficaciousness of future events or higher levels of organization.

In the following pages biological theories and modes of explanation will be examined to see what light they can throw on various controversies concerning the nature of science and the relation of biology to the rest of science. Biology is especially well-suited for this purpose because it is situated between physics and the social sciences both in the traditional arrangement of the sciences and in the stage of its theoretic development. Too often in the past, issues in philosophy of science have been treated either in total abstraction from science or else solely in the context of physical theories. The ensuing discussion will depart from this tradition on both counts. It will concern specific theories, and these theories will be drawn from biology, not physics. Because several highly articulated biological theories exist, our exposition need not rest solely on claims about what is or is not in principle possible. Actual accomplishments can be cited.

In Chapter One, we will discuss the changes that have taken place in genetics, the apparent reduction of modern Mendelian genetics (transmission genetics) to molecular genetics. If this appearance proves to be veridical, then interlevel reduction from one traditional level of science to another is not only possible, it is actual, since a purely biological theory is being reduced to a physico-chemical theory. To make out an intelligent case either for or against this apparent instance of interlevel reduction, both of these theories must be sketched. As interesting as these theories are in their own right, the purpose of this chapter will be to show exactly how much must be done if the reductionist program is to be successful. In Chapter Two, we will turn our attention to evolutionary theory. Here the problem is not so much one of reduction—the situation has not progressed far enough for such a question even to be posed sensibly. Rather it concerns the adequacy of evolutionary theory itself. In Chapter Two we will set out various formulations of evolutionary theory and compare their structures. Because evolutionary theory is still in a state of flux, the conclusions reached in this chapter are by necessity tentative. Some authors have argued that not only evolutionary theory but also all biological theories are suspect. They claim that in biology there are neither scientific theories nor scientific laws. In Chapter Three, we will set out these objections and try to discover their sources. Chapter Four will be devoted to the time-honored problem of teleology. Teleology as a genuine metaphysical position has long ceased to play a role in science; yet biologists continue to use language that sounds vaguely teleological. The question is whether or not this language can be eliminated from biology without loss. If it cannot, there is yet another reason for maintaining that the reductionist program cannot succeed.

My underlying concern in writing this book is to investigate whether there is a single philosophy of science adequate for all science or whether there are many, each appropriate in its own domain. I must admit that I do not address myself openly to this question in the following pages. Instead I deal with specific problems that must be decided prior to making any decision on this main issue. I do not think enough of these problems have been resolved to allow any sort of definitive decision on this larger question. My own prejudices in the matter are mixed. Throughout this book I will urge various positions on a variety of issues without concerning myself too extensively with possible objections to these positions. On this score I must plead lack of space. The exposition is already overly condensed. Nor do I attempt to place these positions on the reductionism-organicism continuum and label myself a reductionist or an organicist. In this book, am I arguing for reductionism or organicism? The answer is neither and both. The slight contributions which I have to make to the reductionism-organicism controversy will be found in Chapter Five.

CHAPTER ONE

The Reduction of Mendelian to Molecular Genetics

REDUCTION AND SCIENTIFIC REVOLUTIONS

The course of science does not run smoothly. Sometimes drastic changes occur. At other times, changes take place gradually in the context of a particular scientific world view. Thomas Kuhn[1] has termed the former phase revolutionary science and the latter normal science. Another distinction related to the preceding distinction is between replacement and reduction (or subsumption). Usually in a scientific revolution, one widely accepted scientific theory is replaced by another. Such changes were accentuated in the early years of science by the replacement of partially supernatural theories like special creation with fully naturalistic theories like Darwin's theory of evolution. During periods of normal science, the changes that take place tend to be less radical. Scientists strive to make their theories as comprehensive as possible, sometimes by subsuming a previously isolated law under a more general law, sometimes by annexing another, less comprehensive theory.

In the past such "reductions" have always occurred within the confines of

[1] *The Structure of Scientific Revolutions,* 2nd ed. (Chicago: University of Chicago Press, 1970).

a single, traditionally defined area of science. During the past two decades, however, something that looks very much like reduction has taken place in genetics, but this time the theories belong to two different scientific domains. Biologists seem to be in the process of reducing Mendelian genetics, a biological theory, to molecular genetics, a theory couched in physico-chemical terms. To many, this transition is significant enough to be dubbed a scientific revolution, even though it might involve only an instance of reduction, not replacement.

As might be expected, neither the distinction between revolutionary and normal science nor the distinction between replacement and reduction is sharp. Although revolutionary periods in science tend to be short compared to the normal periods, any attempt to classify actual events in the history of science results in the construction of a continuum. Some developments in science clearly constitute revolutions—for example, the introduction of evolutionary theory by Darwin and the rediscovery of Mendel's laws. Others are clearly a part of normal science, such as the recognition of the role of mimicry in evolution or the discovery of sex-linked traits. But what of the formulation of the so-called cell and chromosome theories or the discovery of interspecific hybridization? Although certain events in the history of science clearly belong at the far ends of the revolutionary-to-normal continuum, most belong somewhere in the middle.

Similar observations can be made with respect to the replacement-reduction distinction. There is little doubt that Newtonian physics replaced Aristotelian physics, that Lavoisier's oxidation theory replaced the phlogiston theory, and that evolutionary theory replaced special creationist doctrines. At one time in the philosophy of science, examples of clear-cut cases of reduction were just as easy to come by; for example, the subsumption of Kepler's laws of planetary motion and Galileo's law of free fall under Newtonian mechanics and the law of universal gravitation. Such reductions were viewed as homogeneous because the theories involved referred to roughly the same kinds of things and contained the same terms. Frequently, however, reduction takes place between theories concerning phenomena of apparently different kinds, and the theories contain different descriptive terms. The classical example of such a heterogeneous reduction is that of thermodynamics to statistical mechanics. Initially these two theories were developed independently of one another and each made essential reference to entities and processes not mentioned by the other. Thermodynamics was couched in terms of the behavior of macroscopic objects like gases. Statistical mechanics was couched in terms of the behavior of microscopic objects. However, when such key concepts as the temperature of an ideal gas are identified with such mechanical notions as the mean translational kinetic energy of the gas molecules, then formulations very much like the laws of thermodynamics can be derived from the basic postulates of statistical mechanics.

Even if one accepts the preceding analysis of the difference between reduction and replacement, one must recognize the existence of numerous borderline cases, transitions in science that are not total enough to be termed replacements yet are not quite reductions either. Recently, however, several philosophers of science have questioned the distinction itself. They define "reduction" very rigorously and then argue that no event in the history of science actually fulfills the requirements of this definition. Nothing is at the reduction end of the replacement-reduction continuum. This conclusion is sometimes based on doubts about still another distinction—between theoretical and observational terms. Theoretical terms are those descriptive terms occurring in a scientific theory which derive much of their meaning from this occurrence. In isolation from their theory, they mean little or nothing. Any changes in the theory drastically change their meaning. "Electron," "mass," "gene," and "species" are but a few examples of theoretical terms. Observational terms are those descriptive words which are the common property of mankind. Men knew what the words "red" and "mouth" meant long before science emerged as a separate discipline. The names that have been traditionally used to mark this distinction are misleading. The important distinction is not between observational and nonobservational terms but between theoretically committed and theoretically neutral terms. At the outset, theoretical terms usually refer to unobservable entities and processes, but in some cases they can become more or less observable. However, they do not thereby become any less theoretical.

Kuhn, Hanson, and Feyerabend have argued that the theoretical-nontheoretical distinction, like both of the previously mentioned distinctions, is not sharp.[2] The descriptive terms of science can be arrayed on a continuum stretching from those terms which are closely associated with a particular scientific theory to those which are almost theoretically neutral. However, they argue that no descriptive term is completely neutral with respect to all theories. Every descriptive term, to some extent, is theory-laden. The claim that males and females belong in the same species may sound totally unproblematic today—a matter of common sense—but it presupposes a theoretically committed definition of "species," not to mention "male" and "female." The theories involved are too commonplace to mention, but they are no less theories because of their wide acceptance. One consequence of this view of the descriptive terms of science is that reduction in a strict sense becomes impossible. Galileo's law as he formulated it cannot be subsumed

[2] T. Kuhn, *The Structure of Scientific Revolutions,* 2nd ed. (Chicago: University of Chicago Press, 1970); N. R. Hanson, *Patterns of Discovery* (Cambridge: Cambridge University Press, 1958); P. K. Feyerabend, "Explanation, Reduction, and Empiricism," in *Minnesota Studies in the Philosophy of Science,* Vol. III (Minneapolis: University of Minnesota Press, 1962).

under Newton's laws for the simple reason that the two are incommensurable. The most that can be derived from Newton's laws is a formula fairly similar to Galileo's law. Newton and Einstein might both use the word "mass" in their theories, but they do not mean the same thing by this word. Newtonian mass is not the same thing as Einsteinian mass. The net effect of such reflections on the meaning of scientific terms is the obliteration of the distinction between homogeneous and heterogeneous reductions. All become heterogeneous, even when the two theories contain the same terms. The next step in this progression is then to obliterate the replacement-reduction distinction. All changes in science are really replacements. Reduction in any strict sense is impossible.

As appealing as this line of reasoning may seem, it results in all but insurmountable difficulties for the notion of replacement itself. It seems to require that the choice of one scientific theory over another entails an inferential leap which is at least beyond the power of current formulations of logic, if not inherently nonrational. The traditional view is that one theory is preferred to its competitors when it is more comprehensive, more accurate in its predictions, and simpler than its rival theories. For example, Copernicus' system of astronomy is supposedly preferable to that of Ptolemy on all three counts. However, the story is not quite as straightforward as one might wish. According to both theories, the movements of the planets are explained by systems of imaginary circles turning on imaginary circles, termed epicycles. The chief difference between the two theories is that Ptolemy placed the earth near the center of his system; Copernicus the sun. By this maneuver Copernicus was able to make predictions with an accuracy equal to that of Ptolemy's system using somewhat fewer epicycles. In this respect, the increase in simplicity was not great. Rather the chief advantage of the Copernican system was the increased coherence and comprehensiveness of a physical system in which the sun is centrally located and the earth just another planet. For example, the difference in behavior of the superior and inferior planets flowed naturally from Copernican astronomy. When Ptolemaic astronomers tried to account for these differences, they could do so only by the introduction of clearly *ad hoc* devices.

But if all the descriptive terms in the two theories from the most theoretical to the most observational are theory-laden, how can they be compared and one preferred to the other? According to Copernican astronomy, Mercury and Venus should behave differently from the planets outside the earth's orbit. According to Ptolemaic astronomy, there is no reason to expect any such differences, though they can be accommodated once discovered. But if the thesis under discussion is pushed to its logical conclusion, then we are not comparing the "same" phenomena. Ptolemaic Venus is not the same as Copernican Venus. Hence, we have no reason for choosing one theory over another. What has been termed scientific progress is at bottom not entirely a rational affair. If all descriptive terms that occur in a theory are theory-laden and if reduction is interpreted rigorously, then all changes in science are, in varying degrees, instances of replacement and the choice of one scientific theory over another is, to some extent, not a matter of evidence and reason as these notions are now understood. More subtle conceptions of evidence and reason may actually be operative in science, but they have yet to be explicated.

HISTORICAL VS. RATIONAL RECONSTRUCTION IN SCIENCE

The dispute discussed above does not result so much from the disputants' having different views on reduction as from their having different views on the nature of the philosophy of science. The parties on one side view reduction as an historical process concerning scientific theories as they were formulated at the time. Under such a perspective, reduction becomes a complicated, contingent affair replete with inconsistencies and peculiarities. The narrative is rich in detail and highly instructive, but eventually each case becomes unique. The parties on the other side view reduction formally as an investigation into the relations which emerge between two scientific theories after they have been extracted from their historical settings and reconstructed. No obligation is felt to present the two theories exactly as they were formulated by their founders. Rather they are idealized in the manner of textbook expositions. Logical simplicity is promoted but at the expense of historical accuracy. Perhaps all descriptive terms in a scientific theory are theory-laden, but some are certainly less laden than others. The identification of such terms permits the comparison of different scientific theories with minimal falsification. Perhaps no transition in science counts as reduction in an extremely strict sense of the term, but some approach the reductive ideal more closely than others. The transition from one widely held scientific theory to another may well be so traumatic that the protagonists can barely succeed in disagreeing cogently with each other, but not all such changes are equally fundamental. If they were, there would be no difference between revolutionary and normal science.

In this chapter the transition in biology from classical Mendelian genetics to modern molecular genetics will be traced, using both strategies outlined above. We will couch our investigation in terms of later versions of the logical empiricist analysis of reduction and objections to them rather than in terms of some other analysis, because there is no such alternative analysis now current. At the risk of spoiling any surprises that this chapter might hold, let me state at the outset that I find the logical empiricist analysis of reduction inadequate at best, wrong-headed at worst. I must also warn the reader that what follows is not, and is not intended to be, a short history of genetics. Rather it is intended to be a rational reconstruction of the development of two partially independent traditions in genetics. The initial decision as to the appropriate place to draw the line between these traditions was made on historical grounds, but eventually anachronistic modifications had to be introduced if anything like the ideal of formal reduction is to be attained. However, whenever possible, I have noted departures from historical fact. Certain dangers accompany this dual strategy. The most pernicious is the failure to distinguish between the simple, clear-cut relations between two reconstructed scientific theories and the complex relations that exist between successive stages in the development of scientific theories in the course of science. Another is losing track of the philosophical issues at stake in the thicket of historical details that confronts anyone who abandons textbooks to return to the primary material.

CLASSICAL MENDELIAN GENETICS

At the time Darwin published his *Origin of Species* (1859) and Mendel was performing his experiments on garden peas, protoplasm was thought of as the stuff of life; once its complex structure was thoroughly understood, the secret of life would be laid bare. Upon the rediscovery of Mendel's laws at the turn of the century, attention was shifted from protoplasm to the gene. By the early 1930s the structure of the gene was thought to hold the key to the nature of life.

The traditional textbook exposition of Mendelian genetics is reasonably straightforward. It is usually presented as it applies to multicellular organisms that reproduce sexually. Slightly different stories have to be told for other forms of life. Genes, according to this view, are segments of chromosomes, arranged linearly like beads on a string. With some exceptions, chromosomes exist in pairs. The members of these pairs are said to be homologous to each other. Given any pair of homologous chromosomes, one is contributed by the male parent; the other by the female parent. Hence, half of an organism's chromosomes are contributed by one parent and half by the other. Thus, genes also exist in pairs opposite each other in homologous chromosomes. The relative position of a gene in a chromosome is termed its locus. (Sometimes "locus" is also used to refer to the genes themselves.) The alternate forms of a gene occupying the same locus are termed alleles.

At mitosis each chromosome splits longitudinally to form two new chromosomes identical to the original chromosome. When the cell divides, one complete set of chromosomes goes to each of the two new cells. The result of mitosis is two cells identical to the original cell (see Figure 1-1). In this

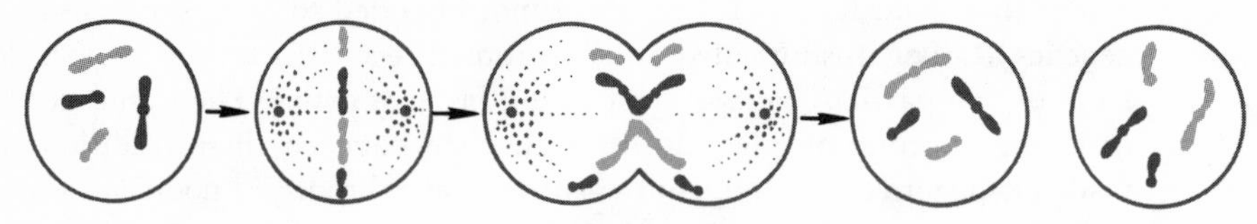

Figure 1-1.
A cell undergoing mitosis to produce two cells identical to the original cell. (Many details have been omitted.)

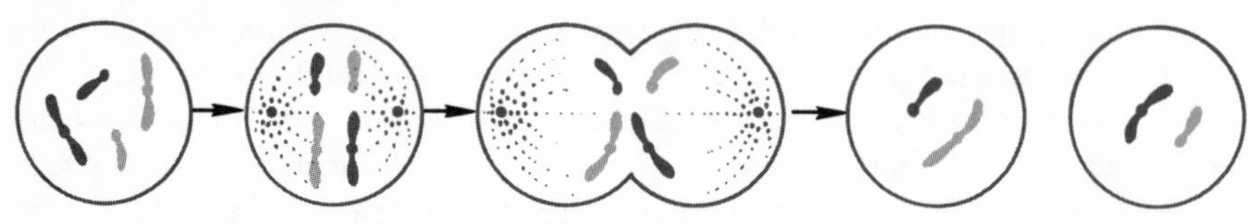

Figure 1-2.
A cell undergoing meiosis to produce two germ cells with half the normal complement of chromosomes.

way multicellular organisms grow and some single-celled organisms reproduce themselves. One consequence of mitosis is that all the cells in the body of a multicellular organism contain the same genetic material, whether they might be nerve, muscle, or bone cells. At some stage between fertilization and gamete formation, however, reduction division (or meiosis) occurs in the gonads of the organism. In meiosis homologous chromosomes line up in pairs. They do not split as they do in mitosis. Rather each chromosome separates from its partner. The two new cells that form (the germ cells) thus contain half the normal complement of chromosomes (see Figure 1-2). At fertilization the male germ cell unites with the female germ cell, reinstating the normal complement of chromosomes in the resulting zygote. A new individual is formed from this zygote by a long series of mitotic divisions.

One important feature of meiosis is that the apportionment of homologous chromosomes to the two germ cells is independent of the history of the chromosomes. One member of each pair goes to each of the two germ cells, but it makes no difference whether the chromosome was originally contributed by the male or by the female parent. Which member of a pair of homologous chromosomes is transmitted to which germ cell is for all intents and purposes purely a matter of chance. Thus, even though we receive half of our chromosomes from one parent and half from the other, the contribution made by each of the grandparents is variable. For example, it is theoretically possible for a child to receive all of the chromosomes contributed by its maternal grandmother and none contributed by its maternal grandfather.

So far chromosomes have been treated as if they were monolithic structures. They are not. Several mechanisms exist for changing the order and combination of genes in a chromosome. The most common are crossing over and recombination. Quite frequently homologous chromosomes twine around each other, break, and then recombine, forming hybrid chromosomes. Thus, even though each chromosome is contributed by a single parent, chromo-

Figure 1-3.
A pair of chromosomes crossing over and recombining during meiosis. (The diagram is greatly simplified.)

somes do not stay pure for very long (see Figure 1-3). Each tends to become a mixture of maternal and paternal genes. If one assumes that genes which are further apart on a chromosome recombine more frequently than those which are closer together, the frequency of recombination can be used to construct a map of the chromosome by following the transmission of characters from generation to generation. For example, genes on opposite ends of a chromosome are exchanged whenever a single crossover occurs, regardless of where it might be, whereas genes that are closer together are separated only if the crossover occurs between them. If it occurs on either side of them, they are exchanged together.

Recombination in turn does not always proceed perfectly. Sometimes homologous chromosomes do not align themselves precisely at meiosis. In such cases, the parts exchanged in recombination are unequal, one chromosome receiving both alleles, the other receiving neither. This process, known as duplication, is one way in which complex chromosomes can be built up from simpler chromosomes. Other mechanisms also exist that change the order of the genes in a chromosome. For example, a single chromosome may twist back on itself, break, and then reassemble so that for part of its length, the order of genes will be inverted. Numerous other chromosomal aberrations also occur. Sometimes a segment of a chromosome will be deleted and then inserted somewhere else in that same chromosome or in some other chromosome. Sometimes whole chromosomes are lost at cell division, resulting in one cell with too many chromosomes and another with too few. In the extreme case, the number of chromosomes in a cell is doubled when the cell fails to divide after chromosomal duplication. In short, almost any chromosomal aberration that could occur does occur.

THE GENOTYPE-PHENOTYPE RELATION

All the preceding discussion was concerned with the genetic makeup of the organism. Nothing was said about the relation between this genetic makeup (or genotype) and the rest of the organism (the phenotype). In general, the position that has predominated in the history of

Mendelian genetics is that the genotype sets limits to the possible variation in the phenotype, limits termed reaction norms. The environment the organism happens to confront determines which of these many possible phenotypes is actually realized. Controversies over the relation between genotype and phenotype have arisen mainly because of certain infelicitous modes of expression employed by some geneticists, especially when the traits under discussion concern such things as human intelligence and insanity. For example, the author of one genetics text can be found saying, "Everything you are, both your physical and mental makeup, is determined by your genetic constitution." As the author makes abundantly clear in later discussion, he does not mean that the genetic constitution is sufficient for any character. Nor is a particular genetic constitution strictly necessary for the production of a particular trait. Usually alternative genetic constitutions can produce the same phenotypes. Rather the relation between genotype and phenotype is variable, much like the relation between smoking and lung cancer.

Although the general nature of the relation between the genotype and environment in producing the phenotype was settled quite early in the history of Mendelian genetics (rare neo-Lamarckians aside), geneticists had little knowledge of the relations between specific genes and the phenotype. In fact, one of the distinguishing features of classical Mendelian genetics is this hiatus in knowledge between the inferred genotype and the finished phenotype. Numerous phenotypic characters were studied in breeding experiments and the genes supposedly responsible for them inferred, but little was known of the specific mechanisms by which the genes acted to produce their respective characters. Contemporary transmission geneticists are still engaged in such an occupation. As late as 1939, C. D. Waddington complained that genetics "was in danger of being considered, by other biologists, as a world of its own, devoted to following the comings and goings of genes whose relevance to other biological phenomena, though uncontrovertible in general theory, could rarely be stated in detail and in particular."[3]

The simplest hypothesis about the relation between genes and characters, and the one that was prevalent in the early days of Mendelian genetics, was the one gene/one character hypothesis. According to this view, each chromosome can be divided unequivocally into a definite number of discrete genes, the phenotype into the same number of equally discrete characters, and a one/one correspondence established between each gene and a single character. Straightforward as this hypothesis seemed to be, numerous conceptual confusions interfered with its being tested and eventually abandoned. The most common confusion was between characters and genes. The ease with which characters could be confused with the genes responsible for them resulted from the fact that the presence, absence, or change of a gene was inferred from the presence, absence, or change of a phenotypic char-

[3] *An Introduction to Modern Genetics* (London: George Allen & Unwin, 1939), p. 7.

acter. To make matters worse, many early geneticists were taken in by the notion of operational definitions—the proposition propounded by P. W. Bridgman in physics and J. B. Watson in psychology. According to the operationist thesis, terms in science should be defined by means of the operations used to investigate their applicability. In the first instance, physical concepts such as length and temperature were to be defined in terms of yardsticks and thermometers, and in the second, psychological concepts such as anger and intelligence were to be defined in terms of behavior and IQ tests. In genetics, the operationist position was interpreted as meaning that genes were to be defined in terms of the breeding experiments from which they were inferred. However, if the operationist position is adhered to rigidly, then the one gene/one character hypothesis becomes an empty tautology. The course of genetics clearly reveals that even the most operationally oriented geneticists did not practice what they preached. There was more to genes than the currently available tests could reveal.[4]

Quite early in the development of genetics, geneticists had to face up to the fact that the phenotype of an organism can be divided into characters in an indefinite number of ways. If one such analysis is to be preferred to another, some justification would have to be given. The problem was that these justifications tended to presuppose a prior knowledge of the genotype, and the genotype in turn had to be inferred from the phenotype. Maybe the phenotype was not divisible unequivocally into discrete characters, but the genotype was divisible into discrete genes. Once a particular gene was isolated, then whatever phenotypic effect it might have could be counted as a character. However, the consequences of the operationist thesis proved to be just as deleterious in genetics as it had been in physics and psychology.

Genes at widely different places in the genome frequently affect the same trait, a phenomenon known as multiple genes. For example, genes at ninety different loci have been discovered which affect eye color in *Drosophila*. On the one gene/one character hypothesis, either all of these genes would have to be considered a single gene, or else eye color in *Drosophila* would have to be considered ninety different traits. Similarly the same gene often affects numerous different traits, a phenomenon known as pleiotropy. For example, alleles at the white eye locus in *Drosophila* affect eye color, the shape of the sperm storage organs in females of the species, as well as several other widely different traits. On the one gene/one character hypothesis, either all of these diverse traits would have to be considered the same trait, or else the gene in question would have to be considered several different genes. The situation was not a happy one. The procedure seemed both logically fallacious and contrary to common sense.

[4] See David Hull, "The Operational Imperative—Sense and Nonsense in Operationism," *Systematic Zoology,* 16 (1968), 438–57.

STRUCTURE, FUNCTION, POSITION, AND THE GENE CONCEPT

The preceding dilemma results from the studious avoidance of theories and theoretical entities in science. Given a realistic interpretation of theoretical entities, the logical snarls can be unraveled. The three criteria most often cited as defining "gene" are position, function, and structure. (The evolutionary dimension of the gene concept necessitated by evolutionary theory will be discussed later.) Is any one of these criteria taken by itself sufficient for two genes being considered instances of the same gene? Is any necessary?

On the usual analysis of the gene concept, neither position nor function is thought of as being sufficient for classing two genes as instances of the same gene, but the two taken together are frequently viewed as being severally necessary and jointly sufficient. No mention is made of structure. On this view, genes which occupy the same locus and affect the same trait are instances of the same gene. If they affect the same trait but differently, they are termed alleles. Numerous different alleles can exist at the same locus, a phenomenon known as multiple alleles (not to be confused with the previously mentioned multiple genes). For example, at the white eye locus in *Drosophila,* fourteen different alleles exist which produce eye color ranging from red to almost pure white. Similarly on this view there is no question but that genes which occupy different loci and affect different traits are different genes.

Hence, it would seem that "gene" is defined solely in terms of position and function. However, geneticists have not been consistent in their treatment of the gene concept, and this inconsistency provides at least circumstantial evidence that other factors are also involved. In the late 1920s geneticists began to suspect that the same gene might function differently at different places in the genome, a phenomenon known as the position effect. Not only did a change in the position of a gene affect its functioning, but also such a change might affect the functioning of other genes as well. For example, there is a gene in *Drosophila* which produces evenly distributed eye pigment when it resides in a chromosomal area which does not stain heavily. When it resides in an area which reacts strongly to the staining process, it produces variegation. On the position or the function criteria, these genes would become different genes when they changed their positions or functions.

Yet geneticists insist on calling these genes the "same" gene. The lengths to which geneticists went to avoid admitting the existence of the position effect attests to the hold which the position and function criteria had on them, but the fact that the position effect appeared to pose such a problem for genetics and could be described coherently within the context of Mendelian genetics indicates that something more fundamental underlay the Mendelian gene concept than just position and function. This something was structure. Genes are theoretical entities in Mendelian genetics. They behave the way they do because of their structure. It is possible for the same gene, structurally defined, to occupy different loci in the genome. It is possible for a gene residing at a single locus to change its function without becoming a different gene. It is possible for a structurally defined gene to be inert, to have no function. All of these phenomena do occur and are described as such by geneticists, indicating that geneticists conceive of genes as specific structures and not as operationally defined in terms of function and/or position. In the following discussion, genes will be treated as structurally defined entities, both because this was the assumption built into the genetic theories of the period and because the distinctions that must be made cannot be made in any other way.

DOMINANCE AND RECESSIVENESS

When the same locus on a pair of homologous chromosomes is occupied by genes with the same structure (i.e., the chromosomes were homozygous for the gene), Mendelians saw no problem concerning the phenotypic character that would result, but when a locus on a pair of homologous chromosomes is occupied by genes with different structures (i.e., the chromosomes were heterozygous for a gene), the phenotypic result can vary. Sometimes one allele completely predominates (the dominant gene) while the other is completely subordinate (the recessive gene). For example, in human beings the gene for brown eyes is dominant to the gene for blue eyes. Here the trait is the visual appearance of the eyes as viewed by other human beings, not the presence or absence of a pigment. In such cases, when two people mate, one homozygous for blue eyes and one homozygous for brown eyes, all of the offspring will be heterozygous for eye color. That is, each possesses one gene of each kind. One is tempted to say that in such crosses all of the children have brown eyes *because* the brown-eye gene is dominant to the blue-eye gene. From an operational point of view, such a causal implication is spurious, because dominant genes are supposedly defined as those genes that behave in this way. From the point of view of genes as theoretical entities with specific structures, the causal implication is appropriate, though the actual mechanisms may be unknown.

The dominance-recessive relation is relative both to the alleles and to the traits in question. An allele can be dominant to one of its alleles but recessive to another. For example, in a particular species of rabbit, the allele for the normal brown coat color is dominant to an allele which results in rabbits which are white except for color around the nose, ears, tail, and feet. This allele in turn is dominant to the allele for albinism. As mentioned previously, alleles at a single locus can affect several different traits (pleiotropy). In such cases, an allele may behave as a recessive for some of the characters which it controls, dominant for others. Hence, no allele is simply dominant or recessive. It is dominant with respect to another allele as far as a particular trait is concerned.

However, sometimes alleles are not simply dominant or recessive to each other, even for a particular trait. For example, if a homozygous white Andalusian fowl is crossed with a homozygous black, all of the offspring will be heterozygous for color, but the observed phenotype will be bluish-gray in appearance, an instance of incomplete dominance. Thus, a continuum exists between complete dominance and complete recessiveness. In addition, sometimes a trait is exhibited more markedly in the heterozygote than in either homozygote, a phenomenon known as overdominance. Sometimes both traits are exhibited in the heterozygote, a phenomenon known as codominance. But the most interesting complication introduced into Mendelian genetics is epistasis. In cases of multiple genes, two or more genes affect the same trait. In epistasis one gene seems to affect another gene. For instance, if two mice are crossed, one homozygous for the dominant color agouti (gray) and one homozygous for the recessive coat color black, all the offspring should be gray. But in such crosses, an occasional albino is produced, and the frequency of the appearance of the albinos is too great to be accounted for by mutation. The explanation for such apparent non-Mendelian ratios is that a second pair of alleles at a second locus also affect coat color. The dominant allele at this locus is for normal coat color; i.e., in the presence of the dominant allele, the color of the coat is determined by the alleles at the primary locus. The recessive allele at the second locus is for albinism. Any mouse which possesses two recessive alleles at this second locus will be albino regardless of the genes at the primary locus. In cases where the mice which are crossed are heterozygous for *both* genes, three-fourths of the offspring should be gray and one-fourth albino. (Neither of the original mice could have been homozygous for the albino gene, since it would have been albino, not gray.) In such cases, the second gene is said to be epistatic to the primary gene and is thought of as somehow "controlling" it. The net effect of all the preceding is that the relation between genes and phenotypic traits is extremely complicated, vastly more complicated than the proponents of the one gene/one character hypothesis had imagined.

CHALLENGES TO MENDELIAN GENETICS

Numerous changes were made during the early years of Mendelian genetics. The principle of independent assortment was restricted just to pairs of genes (or "factors" as Mendel called them) which resided on different pairs of homologous chromosomes. The simple dominant-recessive relation was seen not to be central to Mendelian genetics, as Mendel himself had recognized. But all of these changes were looked upon as modifications of Mendelian genetics rather than as falsifications of it. The two major challenges to Mendelian genetics were made by W. E. Castle and Richard Goldschmidt.

One of the cornerstones of Mendelian genetics was the purity of the heterozygote. When two different homozygotes are crossed, all the offspring are heterozygous. When these offspring are crossed, one quarter of the offspring should be one homozygote, one quarter the other homozygote, and one half still heterozygous. The crucial feature of Mendelian genetics was that the homozygotes would have been unaffected by their passage through the heterozygous state. But what was it that remained pure, the genes or the traits? Mendel had couched nearly his entire exposition in terms of traits. He barely alludes to his genetic "factors." Yet as early geneticists were quick to realize, Mendel's laws apply to only a very few traits if they are interpreted as referring to the phenotype.

Castle (1905) fastened on the latent ambiguity in the genetic literature between genes and traits, arguing that heterozygotes did not remain pure. Instead they "contaminated" each other. In crossing albino and black rats, Castle discovered that the amount of black pigment in the albinos gradually increased with each successive cross. These traits seemed to be mixing because the genes involved were contaminating each other in the heterozygote. Mendelian geneticists attempted to accommodate such cases by introducing additional, nonblending epistatic genes. The gradual increase in pigment in the albinos was due to recombination of the numerous multiple genes responsible for pigment production and distribution, not contamination. Castle was appalled at the introduction of so many *ad hoc* entities. It seemed to him as if Mendelians were piling epicycle upon epicycle in a way which had been so harshly condemned in astronomy. The controversy between Castle and the Mendelians did not last long, depending as it did in so large a measure on the failure to distinguish between genes and traits. Once this distinction was clearly made and the operational predilection of so many early geneticists overcome, even Castle saw that the weight of empirical evidence was against him and he capitulated.

In 1925 A. H. Sturtevant discovered what he termed the position effect. Just as Castle's objections to Mendelian genetics were clouded by the confusion of genes with traits, the controversy surrounding the position effect was complicated by several confusions. For example, numerous quite different phenomena were all classed together under the heading of the position effect. The previously discussed issues concerning the proper definition of "gene" introduced further confusion. The important question was whether the same gene (defined structurally) would have the same phenotypic effect at different places on the same or different chromosomes. Certain data seemed to imply that the function of some genes was affected by their position. If so, then the function of a gene would be determined both by its structure and by its position.

Richard Goldschmidt[5] took the existence of the position effect to be in direct contradiction to the principles of Mendelian genetics and claimed that no amount of reformulation would be adequate to accommodate these recalcitrant data. The whole idea of distinct genes had to be abandoned. The position effect posed two problems for Mendelian genetics. The first, concerning the relation between position and function in defining what is to count as the same or different genes, has already been discussed. The second concerned a mechanism to account for the existence of the position effect. Mendelians were forced to recognize the existence of the position effect but could not explain it. The situation was not unlike that in astronomy with respect to the differences in behavior between the superior and inferior planets. All astronomers recognized that they existed, but according to the Ptolemaic system there was no explanation for these differences. The ancillary devices introduced to account for these differences in the Ptolemaic system were extremely *ad hoc*. But if the sun were in the center of the solar system and the orbit of the earth situated between the inferior and superior planets, the differences flowed naturally from the system, forming an integral part of it.

Gradually it became obvious that further progress in genetics depended on additional knowledge of the fine structure of the gene and of gene action. In classical Mendelian genetics, genes had never been treated as if they were monolithic, because they could mutate, but were they discrete? Were they divisible? If genes were actually arranged on chromosomes like beads on a string, they could be distinguished structurally and recombination could occur only between genes, but if the genetic material was more or less continuous, genes might not be distinguishable by structure alone (a functional criterion might have to be used) and recombination might occur within as well as between genes. Genes mutated, but how? What changes took place? Genes acted in all sorts of ways, but what were the mechanisms involved? Why were some genes dominant, some recessive, and some incompletely dominant? How could one gene "mask" or "control" another gene? Studies in molecular genetics would not only resolve certain conflicts in Mendelian genetics but also answer some of the questions that had gone unanswered. The turning point in the development of molecular genetics was the model of DNA set out in 1953 by J. D. Watson and F. H. Crick.

THE DIVISION BETWEEN MENDELIAN AND MOLECULAR GENETICS

The crucial issue in deciding whether or not Mendelian genetics has been reduced to molecular genetics turns out to be how one distinguishes between the two disciplines in the first place. A careful study of the history of genetics is of some help in making this decision, but as one might expect, there is no point at which Mendelian

[5] *The Material Basis of Evolution* (New Haven: Yale University Press, 1940.)

genetics ends and molecular genetics begins. Although early advances in Mendelian genetics far outdistanced those in molecular genetics, the two disciplines developed simultaneously and interdependently. In this book, we will use Theodosius Dobzhansky's (1970:167) definition of Mendelian genetics. According to Dobzhansky, "Mendelian genetics is concerned with gene differences; the operation employed to discover a gene is hybridization: parents differing in some trait are crossed, and the distribution of the trait in hybrid progeny is observed." Knowledge of the molecular structure of the gene, biosynthetic pathways, decreased enzyme production and such is of no interest to a Mendelian geneticist unless it leads to the discovery of an unsuspected hereditary pattern. Differences that do not affect Mendelian ratios make no difference (see also Simon, 1971).

Molecular genetics, on the other hand, is necessarily a theory of biosynthetic development. Molecular geneticists do not concern themselves with just the molecular structure of the hereditary material and its transmission from one generation to the next. They could not if they wanted to. Rather they are engaged in discovering how molecularly characterized genes produce proteins which in turn combine to form gross phenotypic traits. Quite early, geneticists like C. H. Waddington warned that between genes and gross phenotypic structures, there existed a complex network of chemical reactions. Terms like "dominant" and "epistatic" probably did not refer to reactions between genes but reactions between gene products. Different alleles, for example, might function by producing different amounts of an enzyme.

The difference between Mendelian and molecular genetics is not that Mendelian geneticists treat gross phenotypic traits, whereas molecular geneticists treat molecularly characterized traits. The methods of transmission genetics are as applicable in studying the transmission of a pigment as they are in studying the transmission of a gross phenotypic trait. In fact, most of the initial headway made in deciphering the fine structure of the gene was made by using traditional methods. (See later discussion of the cistron, recon, and so on.) Similarly, the results of molecular studies were frequently read back into Mendelian genetics. For instance, the previously mentioned example of incomplete dominance in Andalusian fowl is better described as a case of codominance, because both genes are fully functional. Both black and white pigments are produced. The gray color depends on the visual acuity of the observer. Brown-eye appearance in human beings is dominant to blue-eye appearance but not because of the production of any blue pigment. The blue appearance of human eyes, like the blue appearance of the sky and clear water, results from the refraction of light and not the presence of a blue pigment.

In retrospect the features of Mendelian genetics that set it apart are the purity of genes in the heterozygote and the absence of any satisfactory explanations for such phenomena as epistasis, pleiotropy, and the position effect. These phenomena were discovered by the traditional methods of

Mendelian genetics and the necessary genes postulated, but the existence of such phenomena and the behavior of their corresponding genes do not flow naturally from the principles of Mendelian genetics in the way that dominance and recessiveness do. The features most characteristic of molecular genetics are the concept of a biosynthetic pathway and the central role of the molecular structure of DNA and RNA in the production of various structural, catalytic, and regulatory proteins. Molecular genetics is more powerful than either classical or contemporary Mendelian genetics because it explains all the types of phenomena explained by them and more. For example, an order of priority exists among epistatic genes in Mendelian genetics. One possible molecular explanation for this type of phenomenon is that these genes occur earlier in the same biosynthetic pathway. Similarly, the position effect in some instances may be due to the existence of operons, multigenic units under the control of a single small segment of DNA (to be discussed later).

Thus, the comparison between Mendelian genetics and molecular genetics, as I have defined them, is asymmetrical. Mendelian genetics is being limited to the analysis of the hereditary material into genes by means of transmission studies and the characterizing of these genes in terms of the resulting Mendelian ratios, whereas molecular genetics is being expanded to include the molecular mechanisms which intervene between DNA and finished proteins. There is no other alternative open. Ideally one should compare two theories of transmission or two developmental theories, but that is impossible. One could compare Mendelian genes to molecular genes, but such a comparison would omit most of both Mendelian and molecular genetics. The only plausible molecular correlate for a dominant Mendelian gene is a highly specified molecular mechanism, not an isolated stretch of DNA. Of course, other reconstructions of Mendelian and molecular genetics are possible. The issue is whether or not any of these reconstructions permit Mendelian genetics to be reduced to molecular genetics in the manner explicated by the logical empiricists.

THE MOLECULAR STRUCTURE OF THE GENETIC MATERIAL

Mendelian geneticists assumed that genes possessed some sort of structure that wholly or partially determined their function, but the nature of this structure was unknown. Several physicists tried to predict what the structure of the gene must be like. E. Schrödinger (1945) thought that genes might be aperiodic crystals. M. Delbrück (1949) thought that they might be submicroscopic steady-state systems. N. Bohr (1932) went so far as to suggest that genes might turn out to be chemically unanalyzable.[6] Hereditary phenomena were too complex to be accounted for by any molecular structure no matter how complicated it might be. In fact, prior

[6] See Blackburn (1966), for the papers by Schrödinger, Delbrück, and Bohr.

to the early 1940s proteins seemed to be the most likely candidates for the genetic material, for only they possessed a richness in structure sufficient to account for hereditary phenomena. The other prime candidate, nucleic acids, were dismissed because their structure was too simple. They were made up of a phosphate, a sugar molecule, and four types of bases arranged linearly. The overall structure of DNA is simple when compared to the structure of various proteins, but simple as it might be, it proved to be adequate to account for hereditary phenomena beyond anyone's most optimistic hopes.

The molecular analog to a Mendelian gene is a segment of deoxyribonucleic acid (DNA). The gross structure of DNA is that of a double helix, a ladderlike molecule twined around itself in a spiral. The sides of the ladder are made up of alternating sugar and phosphate molecules. In fact, it was the presence of phosphorus in nucleic acids and its rarity in proteins that helped to prove that nucleic acid was the hereditary material, because in viral infections and transformation the protein remained outside the cell and only the phosphorus-bearing nucleic acid penetrated. The rungs of the ladder are made up of pairs of four bases—guanine (G), adenine (A), cytosine (C), and thymine (T).[7] A single unit of DNA (a nucleotide) is thus made up of a phosphate-sugar group bonded to one of these four bases. The simplicity of the DNA molecule is enhanced by the fact that each base can bond with only one other base in forming the rungs of the DNA ladder. Guanine can pair only with cytosine, and adenine only with thymine. The requisite complexity is provided by the linear ordering of these base pairs, just as the alphabet can be translated into a series of dots and dashes in Morse code. Hence, a short segment of a DNA molecule can be represented schematically as in Figure 1-4.

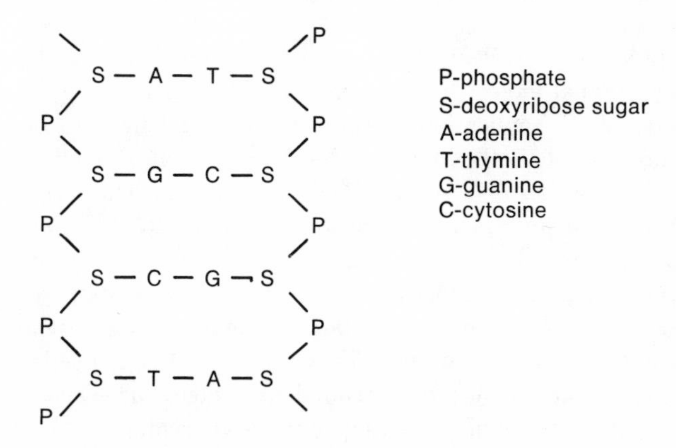

Figure 1-4.
Schematic representation of four nucleotides in a single strand of DNA.

[7] In some viruses and bacteria, cytosine is replaced by methyl and hydroxymethyl cytosine and adenine by methyl adenine.

The functioning of the DNA molecule can be divided into two processes—replication and transcription. Replication is the molecular analog to mitosis. In replication the bonds between the bases which go to make up the rungs of the DNA ladder break, splitting the molecule down the middle. The missing half of each side is then filled with the appropriate nucleotides. Because the two sides of the ladder are complementary, the resulting molecules of DNA will be identical in structure to the original. Half of each new molecule is made of atoms that were physically part of the original molecule and half of new atoms obtained from the environment (see Figure 1-5). The

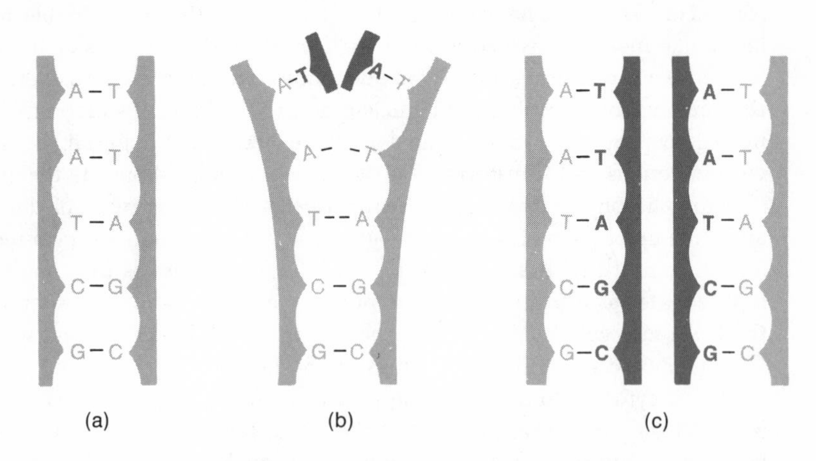

Figure 1-5.
Schematic representation of replication. Light segments represent the original DNA molecule. Dark segments represent new nucleotides added to the original molecule.

end product of replication is two molecules, which have the same structure as the original molecule. (Compare a with c in Figure 1-5). If DNA functioned only in replication, nothing more would have to be said about the organization of the DNA molecules, but DNA also functions in the production of molecules of a second kind of nucleic acid termed ribonucleic acid (RNA) in a process known as transcription. RNA differs from DNA in several respects. It consists of roughly one-half of a molecule of DNA; that is, one side of one ladder. Its single backbone is composed of alternating sugar and phosphate molecules, but the sugar is ribose instead of deoxyribose. Hence, the molecule is termed *R*NA instead of *D*NA. The bases are the same except that uracil (U) or a related base takes the place of thymine. It is through the agency of RNA that proteins are synthesized in a process known as translation.

As complex as proteins are, they are made up of only twenty different amino acids. These amino acids are arrayed linearly in proteins to form polypeptide chains. The order of the amino acids in the polypeptide chain is termed the primary structure of the protein. The complexity of proteins stems from these linear molecules folding back on themselves and combining with one another to form highly complex, three-dimensional molecules. However, these higher levels of organization are almost exclusively a function of the primary structure. As always, of course, the environment does affect the genotype-phenotype relation.

The problem in discovering the genetic code was to establish a relationship between the base pairs of DNA (the letters in the genetic code) and the amino acids of proteins through the mediation of the base pairs in RNA. Which and how many bases designate which amino acids? The letters of the genetic code had to be arranged into words termed codons. The question was how many letters constitute a codon and which codons code for which amino acids? The answer was again simpler than expected. If all codons are of the same length, the minimal length of a codon is three letters. If codons are only one letter long, only four amino acids can be stipulated by the four bases A, U, G and C. If codons are only two letters long, only sixteen codons will be formed. When the length of a codon is expanded to three letters, sixty-four codons are possible, more than enough to code for the twenty amino acids. At this juncture one might reexamine the assumption that all codons are of the same length (an exceedingly unattractive alternative) or else attempt to establish a relation between three-letter codons and the twenty amino acids. The correspondence cannot be one/one. Some codons may code for no amino acid at all (nonsense). Some may serve to initiate or terminate developmental sequences (punctuation). More than one codon may code for the same amino acid (degeneracy or redundancy). And if God is really unkind, the same codon may code for more than one amino acid (ambiguity).

As it turns out, the code is triplet and all but one of the preceding possibilities obtain. Amino acids are stipulated by two to six different codons; for example, the amino acid phenylalanine is specified by only two codons (UUU and UUC), while the amino acid leucine is specified by six codons (UUA, UUG, CUU, CUA, CUC, and CUG). Some codons serve as terminating punctuation (UAG, UAA, and UGA). Because these codons do not code for any amino acid, they are sometimes termed misleadingly "nonsense." At one time, the code appeared to be ambiguous. For example, GGA was thought to code for both glycine and glutamic acid. Such ambiguities have now been successfully eliminated. Thus, although the genetic code is not as orderly as one might wish, it is still remarkably simple.

As we mentioned previously, DNA serves two functions—in replication it is duplicated and in transcription it produces complementary molecules of RNA. In both processes the order of the bases in the original molecule is retained in the subsequent molecules. The molecules of RNA in turn function, with the aid of certain proteins, to produce other proteins in a process known as translation. In translation, at last the sequence of bases coded in the DNA is converted into sequences of amino acids to form proteins. There are three different kinds of RNA. One kind of RNA along with certain proteins goes to form small bodies in the cytoplasm which function as seats of biochemical activity. These cytoplasmic bodies are termed ribosomes, and the RNA they contain is termed ribosomal RNA (rRNA). Molecules of a second kind of RNA, messenger RNA (mRNA), align themselves on one or more ribosomes in preparation for the production of proteins. Molecules of transfer RNA (tRNA) then serve to transport the appropriate amino acids to the appropriate places on the mRNA.

The structure of tRNA requires some explanation, because it plays such a central role in protein synthesis. Each molecule of tRNA is shaped like a clover leaf. Three bases are exposed at the middle lobe and are free to pair with the appropriate codon of the mRNA. These three-base triplets on

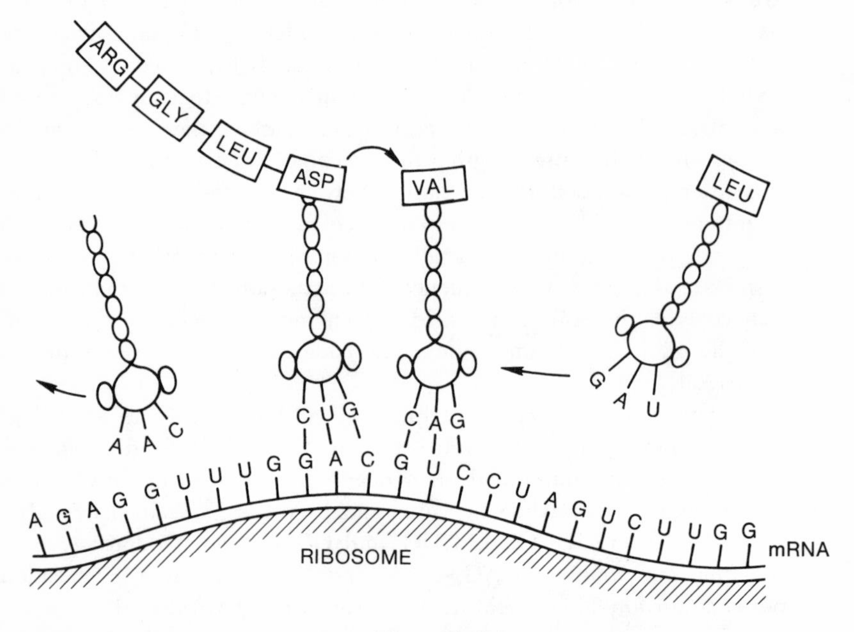

Figure 1-6.

In the above diagram, four molecules of tRNA are shown functioning in protein synthesis. The molecule on the right, charged with leucine is about to pair with the appropriate codon on the mRNA. The second molecule of tRNA charged with valine has paired with its codon and the growing polypeptide chain is about to be transferred to it. The transfer has taken place for the third and fourth molecules of tRNA. The fourth molecule of tRNA has been stripped of its amino acid and discharged. The process continues in this step-by-step fashion until the entire message have been translated.

the middle lobe of molecules of tRNA are termed anticodons, because they are complementary to the codon triplets of the mRNA. In protein synthesis, the appropriate amino acid is attached to the stem of the appropriate molecule of tRNA opposite the middle lobe by means of adenosine triphosphate (ATP) and several enzymes. Such charged molecules are then termed aminoacyl-tRNA. Next, the charged molecules of tRNA become associated with the appropriate codons on the mRNA. The codons in turn on the mRNA are read sequentially, the ribosome moving relative to the mRNA much as a punched tape moves relative to a computer head. As each codon passes, the appropriate aminoacyl-tRNA enters, becomes attached to the growing polypeptide chain, and then leaves empty, having transferred the polypeptide chain to the next aminoacyl-tRNA (see Figure 1-6). This process is not only precise but also extremely intricate, involving as it does mRNA, ribosomes, aminoacyl-tRNA, several enzymes and an activating molecule of guanosine triphospate (GTP).

BIOSYNTHETIC PATHWAYS

DNA replication, transcription to RNA, translation to protein, and the genetic code go together to form one of the most important constituents of molecular genetics. Another is the central role played by biosynthetic pathways. One of the major lacunae in biology prior to the emergence of molecular genetics was between genes and gross phenotypic characters. In Mendelian genetics, though some modifications were introduced, the assumption was that the connection was fairly direct. As molecular biology progressed, the extent of the complexity in the production of a single gross character, like eye color in *Drosophila,* was realized. Schematically what happens in a biosynthetic pathway is that a particular gene produces via RNA an enzyme that acts on a substrate to produce a product of some kind. This product in turn serves as a substrate for a second enzyme produced by a second gene to synthesize a second product, and so on (see Figure 1-7).

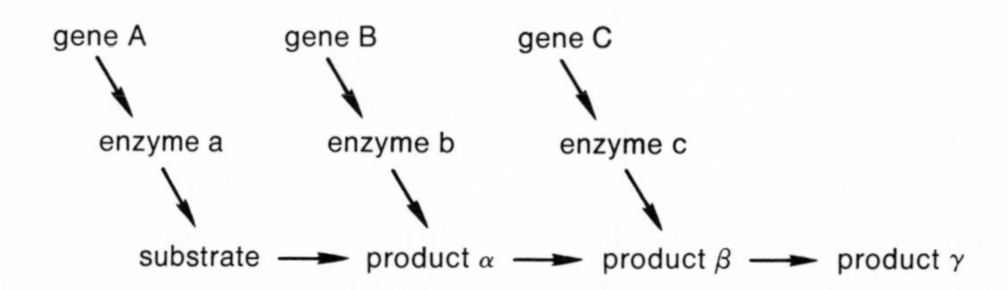

Figure 1-7.
A diagram of a biosynthetic pathway converting a substrate into an end product through the successive action of three different genes.

Although for short segments, biosynthetic pathways may be simple chain-like affairs, in general they form complex branching, cyclical nets. One metabolite may serve as a substrate for more than one reaction. Conversely, the same metabolite may be produced by several alternative pathways. Such redundancy is of prime evolutionary significance at both the molecular and the macroscopic levels. One of the most common ways of deciding which factors influence a particular causal chain is to vary one at a time to see what changes are produced. If the elimination of *A* results in an elimination of *B,* then *A* must have been a necessary condition for *B.* This strategy is often frustrated by the frequency with which a biologically important function can be performed in alternative ways. For example, sap must rise in deciduous trees in the spring if they are to survive and female rabbits must ovulate if the species is to persist. But when these processes were investigated, no one feature could be isolated that was absolutely necessary. In deciduous trees, if the roots were cut off and the tree set in a container of sap, the sap would rise. Hence, root pressure was not a necessary condition in the rise of sap. If another tree were stripped of its leaves, the sap still rose. Hence, leaves performed no necessary function in the process. If a circle of phloem cells were killed around the base of the tree, the sap still rose, and so on. A comparable story can be told for ovulation in rabbits. In each case, the vital functions were overdetermined. Any one of several alternative and reinforcing mechanisms were available to perform them. Such redundancy is extremely adaptive in evolution. Whenever a process can be performed in only a single way, an inhibition of that process frequently will lead to the death of the individual and possibly the eventual extinction of the species. For example, somewhere in his evolutionary development, man lost the ability to synthesize vitamin C. If it had not been available to him in his diet, this mutation would have become established in man only at the cost of his extinction. Redundancy, of course, also has its evolutionary price, but where it exists it is counterbalanced by the cost of its alternative. (See Chapter Four on functional explanation.)

THEORY REDUCTION IN GENETICS

Now that the outlines of classical Mendelian genetics and molecular genetics have been sketched, we are in a position to address ourselves to the question of reduction in greater detail. Several philosophers have presented semiformal analyses of the relation that exists between two theories when one is "reduced" to another. For example, Kemeny and Oppenheim[8] describe what they term "indirect reduction." However, their analysis is actually of what

[8] J. G. Kemeny and P. Oppenheim, "On Reduction," *Philosophical Studies,* 7 (1956), 6–17.

we have termed "replacement." For example, both the theories of phlogiston and oxidation attempted to account for the phenomena of heat. Eventually the phlogiston theory was abandoned for the oxidation theory. The phlogiston theory was in no sense "reduced" to the oxidation theory. The Kemeny-Oppenheim reconstruction is useful, however, as an analysis of theory replacement. According to these authors, if one theory is preferred to another, the preferable theory must be more highly systematized, have greater scope, or produce more accurate predictions than the other. Two additional criteria that are more problematic, though crucial, are simplicity and compatibility with other well-established theories.

In the classic example of replacement, the decision between Ptolemaic and Copernican astronomy was not completely one-sided. As calculation devices, both theories were well systematized and permitted predictions concerning the positions of the heavenly bodies with equal accuracy when worked out in equal detail. The only difference with respect to such predictions is that Ptolemaic astronomy required a few more epicycles than Copernicus' system if predictions of equal accuracy were to be made. As physical systems, Copernican astronomy was superior to Ptolemaic astronomy. It explained various phenomena that Ptolemaic astronomy accounted for in only a casual, *ad hoc* manner—primarily differences in the behavior of the inferior planets (Venus and Mercury) and the other planets. The Copernican system also had its problems. Stellar parallax should have been observable, but it was not. The greatest hurdle for Copernican astronomy, however, was its incompatibility with all the rest of Aristotelian science from the physics of natural place to the four-humor theory of disease.

The classic analysis of reduction is that of Woodger, Nagel, and Quine,[9] an analysis that has been termed direct reduction. According to this view of reduction, both the reduced theory and the reducing theory are thought of as being essentially correct in their respective domains. The reducing theory, however, is of broader scope than the reduced theory and perhaps provides more accurate predictions in the domain of the reduced theory. When the two theories employ the same descriptive terms, the reduction is termed homogeneous; if not, it is termed heterogeneous. In heterogeneous reduction, all of the primitive terms of the reduced theory (both the names of classes of individuals and predicate terms) must be associated with appropriate terms in the reducing theory by means of reduction functions. The reducing theory may also include additional descriptive terms not associated directly with any of the terms in the reduced theory. These reduction functions are looked upon either as physical hypotheses expressing correlations or

[9] J. H. Woodger, *Biology and Language* (Cambridge: Cambridge University Press, 1952); E. Nagel, *The Structure of Science* (New York: Harcourt, Brace, & World, 1961); W. V. Quine, "Ontological Reduction and the World of Numbers," *Journal of Philosophy,* 61 (1964), 97–102.

as correspondence rules. In either case, they are synthetic claims and require empirical support. Finally, from a conjunction of the reducing theory and these reduction functions, the reduced theory must be deducible.

Popper, Feyerabend, and Kuhn[10] have argued persuasively against the possibility of direct reduction in science. Even though the same words may appear in two different theories in homogeneous reduction, this does not mean that these terms are being used in the same sense in the two theories. In heterogeneous reduction, the situation is even more marked, because the terms themselves differ. Additional doubts have been raised as to the sense in which two different theories in direct reduction can be said to explain the "same" phenomena, if one grants that even observational terms are theory-laden. In Aristotelian physics, the path of a heavenly body calls for explanation when it departs from perfect circularity. In Newtonian physics, departures from perfect rectilinearity must be explained. Yet, viewed differently, Popper, Feyerabend, and Kuhn have provided the best analysis to date of reduction. In order to reduce an earlier, less adequate theory to a later, more adequate theory, the reduced theory must be corrected. The theories being related are not purely historical entities but logical reconstructions. If both historical accuracy and logical precision are demanded, then formal reduction is impossible. As Kenneth Schaffner[11] has argued, if strict historical accuracy is sacrificed somewhat to allow for the correction of the reduced theory, then the Popper-Feyerabend-Kuhn thesis of antireduction can be converted into a general reduction paradigm. This is all that can be meant by formal theory reduction. The demand for more is mistaken.

According to Schaffner's general reduction paradigm, five requirements must be met in any case of reduction:

1. All of the primitive terms in the corrected version of the theory being reduced must appear in the reducing theory or be associated with one or more of the terms in the reducing theory. These associations will be of referential identity; that is, both terms will denote the same entities though they may not mean the same thing. Each term will have a primary meaning in its own theory and a secondary meaning conferred on it by its appearing in the reduction function.
2. From the reducing theory in conjunction with the reduction functions, the corrected version of the reduced theory must be deducible.
3. The corrected version of the reduced theory must not only correct the original theory but also indicate why it was incorrect.
4. The original theory being reduced must be explicable in a loose, informal sense by the reducing theory.

[10] Karl Popper, *Conjectures and Refutations* (New York: Basic Books, 1962); P. K. Feyerabend, "Explanation, Reduction, and Empiricism," *Minnesota Studies in the Philosophy of Science,* Vol. III (Minneapolis: University of Minnesota Press, 1962); Thomas Kuhn, *The Structure of Scientific Revolutions* (Chicago: University of Chicago Press, 1970).

[11] "Approaches to Reduction," *Philosophy of Science,* 34 (1967), 137–47; see also Schaffner's "The Watson-Crick Model and Reductionism," *British Journal for the Philosophy of Science,* 20 (1969), 325–48.

5. The original theory and the corrected theory must be strongly analogous to each other.

Within Schaffner's analysis, the explication of any case of reduction requires the specification of three different theories—the original theory, the corrected theory, and the reducing theory. In the case of genetics, the original theory is the textbook version of classical Mendelian genetics set out in the early pages of this chapter, the corrected theory is initially modern transmission genetics, and the reducing theory is modern molecular genetics. Classical Mendelian genetics, as we have characterized it, concerns the discovery of Mendelian ratios in the transmission of observable phenotypic characters from generation to generation and the postulation of the requisite number of primary and epistatic genes. The characters are of the "eyes look blue" variety. Genes are viewed as beads on a string, discrete entities, structurally defined, though functionally determined, which control phenotypic traits either directly or indirectly. Crossover occurs only between genes, and mutation is some change in the structure of the gene. Hence, the units of function, crossover, and mutation are coextensive. This view of Mendelian genetics is, of course, distorted. We have set up a "straw gene" so to speak. Early geneticists took the "beads on a string" metaphor no more seriously than physicists took Bohr's miniature solar system model of the atom. Both models merely provided some plausible antecedent properties for genes and atoms respectively, properties that were not formally part of the theories being developed. The eventual success of a model stems in a large measure from the disanalogies that exist between the model and the natural phenomena it is supposed to represent. The ways that atoms differ from star systems and chromosomes from strings of beads proved to be the important features in the future development of atomic and gene theories. (See later discussion of analogies in Chapter Four.)

Transmission genetics is looked upon as the modern descendant of classical Mendelian genetics because it retains the techniques and modes of explanation of Mendelian genetics, but the two also differ significantly. The differences result from a shift in the type of character that transmission geneticists now trace. No longer do geneticists limit themselves just to the "eyes look blue" type of phenotypic trait. They now distinguish between perceptually identical characters on the basis of biochemical differences. Furthermore, the evolution of classical Mendelian genetics into modern transmission genetics was not independent of the emergence of molecular genetics. Improvements in transmission genetics facilitated biochemical research, and discoveries concerning the biochemical nature of gross phenotypic characters permitted transmission geneticists to refine their analyses. As Sonneborn (1963: 22–26, in Appleton, 1970) has rightly observed:

> To some extent, genetics has been biochemical and molecular almost from the start. Nearly 60 years ago, Garrod pointed out the existence of gene-controlled

biochemical traits in man. There has never been a time since then when comparable studies and theoretical constructs were not current. Second, some of the most spectacular triumphs of the new genetics—such as Benzer's revelations of the number and array of subunits in a gene—were achieved by classical genetic methodology.

Classical Mendelian genetics and modern transmission genetics share a common methodology and mode of explanation. They differ only with respect to the type of characters they treat. Because of the advances made in molecular biology, transmission geneticists now frequently trace the transmission of chemically characterized phenotypic traits. Two specimens may look blue, but one might look blue because it contains a blue pigment, whereas the other might look blue because of light refraction. Molecular genetics differs from both classical Mendelian genetics and modern transmission genetics in its concern with the molecular structure of the genetic material and the biosynthetic pathways that eventuate in molecularly characterized phenotypic traits. Thus, if the requirements of theory reduction are to be fulfilled, the primitive terms of transmission genetics must be connected with those of molecular genetics by the appropriate reduction functions.

Schaffner distinguishes between two types of terms for which reduction functions must be provided—class terms and predicate terms. In transmission genetics, the relevant classes of individuals are genes. Every gene in transmission genetics must be identified with one or more segments of DNA or RNA. Similarly, all the predicates of transmission genetics like homozygous, dominant, recessive, codominant, and epistatic must be effectively associated with open sentences in molecular genetics. Initially the reductionist program seems quite plausible for genetics. Genes could be identified with specific segments of DNA, epistasis could be explained in terms of sequential and convergent biosynthetic pathways, pleiotropy could be accounted for by diverging pathways, position effect in terms of higher levels of organization in the genome such as operons, and so on. However, when one attempts to flesh out this impressionistic first approximation with specific reduction functions, the story gets quite messy and considerable doubt is cast on the sensibleness of the reductionist program.

REDUCTION FUNCTIONS — THE GENE

In the preceding, admittedly simplified, characterization of classical Mendelian genetics, genes were treated as if they were discrete and monolithic structures. The units of crossover, mutation, and function were supposed to be identical. They are not (see Muller, 1927; in Peters, 1959). The genetic material is more or less continuous, being made up of a series of nucleotides. The smallest units of crossover and mutation turn out to be a single nucleotide. Point mutations consist in the change of a single base pair. For example, a DNA anticodon which transcribes a particular segment of mRNA might be changed from CAT to CTT. The resulting segment of mRNA would thus be changed from GUA to GAA and the corresponding amino acid from valine to arginine. The change might effect the functioning of the protein significantly; it might not. Because the genetic code is redundant, certain mutations will have little or no effect on the resulting proteins; for example, a change from UUU to UUC would still code for phenylalanine. The slight changes that might occur would result from the fact that different tRNAs would be required and not all types of tRNA are equally abundant in different tissues of different species of organisms. Regardless of the degree of the phenotypic effect that might or might not occur, such substitutions would still count as mutations, because the structure of the DNA segment would be altered. Nor can crossover alone be used to delineate functional units. Crossover can occur between any two nucleotides. The sugar-phosphate bonds within functional units do not differ from those between them.

If modern transmission genetics is to be reduced to molecular genetics, then Mendelian genes must be identified with certain segments of DNA or RNA. Schaffner, following the lead of most biologists since Benzer (1955, in Peters, 1959), has suggested identifying the Mendelian gene with the cistron, a functional unit delineated by the *cis-trans* test. Neither mutation alone nor crossover alone can be used to distinguish functional units of DNA, because the smallest unit of each is a single nucleotide. However, if the effects of both are followed simultaneously, functional units can be determined in many if not in all cases. This is precisely the point of the *cis-trans* test. If two mutations reside on the same homologous chromosome, they are said to be in the *cis* position. If they reside on opposite homologous chromosomes, they are said to be in the *trans* position. If crossover occurs between these homologous chromosomes, two mutations in the *cis* position can come to occupy the *trans* position, and vice versa (see Figure 1-8). If the two

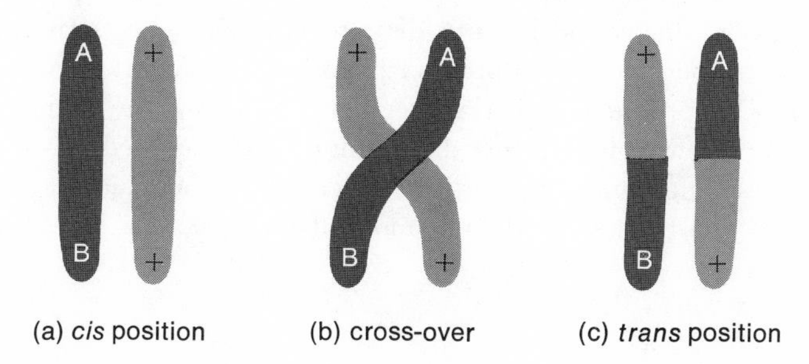

Figure 1-8.
A diagram of a pair of homologous chromosomes with mutations in the *cis* position crossing over to produce chromosomes with mutations in the *trans* position. + stands for the normal allele; A and B for the mutations.

mutations reside in different functional units on the chromosomes, then it makes no difference whether they exist in the *cis* or the *trans* positions; the two units will function regardless. But if the two mutations occur within the same functional unit, the *cis* form may well behave differently from the *trans* form. In the *cis* position, at least one segment in one chromosome is unaffected and will function normally. In the *trans* position, both chromosomes contain a mutated gene in the same functional unit, and neither segment of DNA may function normally. Thus, by definition, any two mutations that exhibit the *cis-trans* effect belong to the same cistron. Those that do not probably do not belong to the same cistron.[12]

The identification of the Mendelian gene with the cistron will not do for several reasons. First and foremost, as Ernst Mayr and Michael Ruse have pointed out, "cistron" is not a term from molecular genetics; it belongs to modern transmission genetics, depending as it does on crossover and recombination studies. No knowledge of the molecular structure of DNA or biosynthetic pathways is involved. Though "cistron" verges on being a word from molecular genetics and was instrumental in the development of this new theory, it does not quite make the grade. Some indirect support for this claim can be gathered from the fact that the *cis-trans* test and its associated concept are becoming less and less important in genetics, as indicated by the decreasing frequency with which they are used in the literature. However, even if the cistron were counted as a term from molecular genetics, Schaffner's identification still would not do because of the existence of various levels of organization in the DNA molecule which the *cis-trans* test fails to discern properly, and because the results of the *cis-trans* test do not always agree with those of other tests designed to delimit functional units in the genetic material.

The identification of the Mendelian gene with segments of DNA that are responsible for the production of a single molecular product (a molecule of RNA) seems more promising. The one gene/one character hypothesis foundered on the impossibility of analyzing the gross phenotype of living organisms unequivocally into discrete unit characters. In this respect, the one gene/one molecular product hypothesis is on surer ground, because the molecular products of various segments of DNA are themselves reasonably discontinuous. Hence, this discontinuity can be used to divide the DNA molecule into units. In addition, such structural clues as terminator anticodons can be used to delimit molecular genes.

12 The explanation of the *cis-trans* test given in the text is applicable only to those organisms that contain homologous chromosomes, but a similar story can be told for those that do not. All that is necessary for the *cis-trans* test is that the two mutated genes come to coexist in the same cell or medium so that complementation of function can be tested.

However, even this identification is too simple. Certain segments of DNA exist which do not themselves code for any amino acid, yet function in the overall genic system. Sometimes the functioning of one or more cistrons is controlled by additional segments of DNA, a regulatory gene and an operator. The regulatory gene produces a protein via the requisite RNA molecules. This protein resides on a segment of DNA termed the operator. When the protein is attached to the operator, it inhibits the functioning of the adjacent cistrons. When the concentration of the substances produced by the adjacent cistrons falls too low, the inhibiting protein is released and then these cistrons begin to produce their respective substances until the requisite concentrations are reached again, whereupon the inhibitor attaches itself to the operator again. The operator in conjunction with the adjacent cistrons which it controls is termed an operon. Any adequate reduction function for "gene" must at least be able to accommodate the existence of operators which code for no amino acid but which still are operative in the functioning of other segments of DNA which do. Perhaps a reasonably simple reduction function for "gene" can be formulated, but because of the features of genic action just described as well as others, this reduction function is likely to express a disjunctive, one–many relation. That is, the notion of a Mendelian gene will be associated with several alternative molecular structures. Such an identification will do for the purposes of reduction, however, because the inferences in reduction must be from the presence of a molecular structure to the presence of a Mendelian gene, not vice versa. If the inferences in reduction had to be from statements about Mendelian genes to specific molecular structures, such an identification would be inadequate. To use the words introduced for discussing the genetic code, redundancy is accceptable for reduction, but not ambiguity.

REDUCTION FUNCTIONS — PREDICATE TERMS

In Mendelian genetics, numerous different characteristics and behaviors are predicated of Mendelian genes. A certain small portion of these predicates pose no real problems for the reduction of Mendelian genetics to molecular genetics. These are the purely structural terms like "homozygous" and "heterozygous." For example, to say that two organisms are homozygous for a trait is to say that they possess genes residing opposite each other on homologous chromosomes which have the same molecular structure. If they are heterozygous for the trait, the genes have different molecular structures. The only complication that arises in these identifications stems from the redundancy of the genetic code. The temptation to go one step further and try to distinguish between those amino acid substitutions that significantly alter the functioning of the resulting proteins from those that do not would introduce unnecessarily prohibitive complexities at this stage of the analysis. One should also note that the notion of homologous chromosomes is assumed in this identification. Homologous chromosomes are those chromosomes that pair at meiosis, a fact that can be derived from cellular cytology.

When we turn from such purely structural features of genes to their functioning in the production of phenotypic characters, the situation rapidly becomes more complex. To say in Mendelian genetics that one allele is "dominant" to another with respect to a particular trait is to say something about how the character which these alleles control is transmitted from one generation to the next. These characters in turn are usually gross phenotypic characters, often described in terms of human perceptions. The only philosopher of science who has addressed himself to the requisite reduction functions for Mendelian predicate terms is Kenneth Schaffner. His identification of the Mendelian gene with the cistron has already been discussed. Although this suggestion turns out to be inadequate, it is a good start. His proposed reduction function for "dominant," however, is extremely unsatisfactory. Schaffner suggests associating the predicate word "dominant" with the open sentence "x is capable of directing the synthesis of an active enzyme" such that "$gene_1$ is dominant" when and only when "DNA $segment_1$ is capable of directing the synthesis of an active enzyme." In this instance, Schaffner's reduction function is of the right kind. It refers to molecular structures and their functioning in the production of proteins. But it misrepresents the Mendelian notion of dominance. In the first place, many genes that behave in a dominant fashion with respect to their alleles have nothing to do with the production of enzymes; for example, those genes that code for the mRNA of structural proteins. Second, "dominance" is a relative notion in Mendelian genetics. No gene is just dominant. It is dominant with respect to certain of its alleles for certain traits in a range of environments. It may be recessive with respect to other alleles or other traits or in other environments. Almost all of the genes that code for specific enzymes produce enzymes that are active to some extent. Hence, according to Schaffner's analysis, all of these genes would have to be termed "dominant," even though most of them are recessive to one or more of their alleles.

If Mendelian genetics is to be reduced to molecular genetics, then better reduction functions than these will have to be formulated. However, the formulation of such reduction functions turns out to be a good deal more formidable than one might at first expect. There are several dimensions to the problem. First, gross phenotypic traits must be translated into molecularly characterized traits. The methods of Mendelian genetics can handle gross phenotypic traits as well as molecularly characterized traits, but the former are too crude for the purposes of molecular genetics. Second, various kinds of Mendelian predicate terms (for example, "dominant") must be associated with one or more types of molecular mechanisms (for example, production of an active enzyme). As mentioned with respect to possible reduction functions for the gene, these correlations between molecular and Mendelian terms may be one–one or many–one, but they cannot be one–many. A single molecular mechanism cannot be associated with several different Mendelian predicate terms. Third, after all the changes in Mendelian genetics necessary for the reduction have been made, this "corrected" version of Mendelian genetics must remain strongly analogous to the original formulation of Mendelian genetics. For example, in traditional Mendelian genetics, there exists a certain symmetry between dominance and recessiveness as well as a certain order of priority in series of epistatic genes. These two relations are combined in cases of dominant and recessive epistasis. One would expect these relations to be retained more or less in the corrected versions of Mendelian genetics, which in turn might lead one to expect them to be reflected somewhat in the corresponding molecular mechanisms. For instance, one might expect dominant and recessive epistasis to be produced by a combination of those molecular mechanisms that produce epistasis with those that produce dominance and recessiveness.

One does not have to look very deeply into the relation between Mendelian and molecular genetics to discover how naive the preceding expectations actually are. Even if all gross phenotypic traits are translated into molecularly characterized traits, the relation between Mendelian and molecular predicate terms express prohibitively complex, many–many relations. Phenomena characterized by a single Mendelian predicate term can be produced by several different types of molecular mechanisms. Hence, any possible reduction will be complex. Conversely, the same types of molecular mechanism can produce phenomena that must be characterized by different Mendelian predicate terms. Hence, reduction is impossible. Perhaps these latter ambiguities can be eliminated by further "correcting" Mendelian genetics, but then we are presented with the problem of justifying the claim that the end result of all this reformulation is reduction and not replacement. Perhaps something is being reduced to molecular genetics once all these changes have been made in Mendelian genetics, but it is not Mendelian genetics.

MENDELIAN PREDICATE TERMS AND MOLECULAR MECHANISMS

The first step in the argument set out above is to show that gross phenotypic characters will not do for the purposes of molecular genetics. In Mendelian genetics, two alleles are termed incompletely dominant if the heterozygote exhibits a trait that is intermediate in expression between the two homozygotes. At the molecular level, however, additional distinctions must be made. The example given previously of incomplete dominance was the crossing of a black Andalusian fowl with a white Andalusian fowl to produce steel-gray offspring. This intermediate trait could result from a variety of molecular situations. Perhaps a blue-gray pigment is being produced. Perhaps only a black pigment is being produced, but in reduced concentrations. Perhaps both black and white pigments are being produced in equal amounts. These pigments could be distributed evenly throughout the feathers or gathered together in small patches. In the latter case, the gray appearance would be a function of the visual acuity of the observer. Depending on which state of affairs actually existed, the trait would have to be assigned to different Mendelian categories. If only a black pigment is being produced but at reduced concentrations, the transmission pattern would be appropriately termed incomplete dominance. If both black and white pigments were being produced, then it would be an example of codominance. In point of fact, color transmission in Andalusian fowl turns out to be an instance of codominance, not incomplete dominance. Both alleles are completely operative, producing both black and white pigments which exist separately in small patches on the feathers.

The message of the preceding example (and many others could be given if space permitted) is that the reduction of Mendelian genetics to molecular genetics requires the substitution of molecularly characterized traits for the traditional traits of Mendelian genetics, and such substitutions entail extensive reclassification. For example, numerous phenomena are now explained in Mendelian genetics in terms of recessive epistasis; for example, coat color in mice, feather color in Plymouth Rock and Leghorn chickens, a certain type of deaf-mutism in man, feathered shanks in chickens, and so on. There is little likelihood that all of these phenomena are produced by a single type of molecular mechanism. At best, several alternative mechanisms are involved. Conversely, there is little likelihood that these various molecular mechanisms always produce phenomena which are appropriately characterized by the Mendelian predicate term "recessive epistasis."

One might counter that, of course, molecular genetics cannot concern itself with such gross characters as "eyes look blue." All such traits will have to be translated into molecular terms like "presence of a blue pigment," or some such. However, comparable problems exist at the molecular level. For example, several different molecular mechanisms can produce molecular phenomena something like "incomplete dominance," and it seems likely that these mechanisms can produce other phenomena as well. For example, assume that the steel-gray color in Andalusian fowl resulted from a decrease in the production of black pigment. The black pigment is just as black, but it exists in a decreased concentration. This decrease could have several sources. Perhaps the concentration of the enzyme necessary for the reaction has been cut in half. Perhaps only half as many mRNA molecules coding for the structure of the black pigment are being produced. Perhaps the concentrations of the enzyme and the structural mRNA are unchanged, but the enzyme is less effective as a catalyst. The latter could result from a slight change in the structure of the enzyme or the structural mRNA. And so on.

Conversely, the same molecular mechanisms can produce different phenotypic effects. For example, in a heterozygote, one allele may be completely operative, the other completely inoperative. Depending on the nature of the alleles and the biosynthetic pathways in which they are functioning the effect can vary. It may be the case that the single allele can produce all the product necessary to maintain the reaction at full capacity. If so, the phenomenon would be termed "dominant." Or it might be the case that the presence of the single operative allele decreases the rate of the reaction only slightly or perhaps cuts it in half. In such cases, the phenomenon would be termed "incomplete dominance." Hence, it would seem that even at the molecular level the relation between Mendelian predicate terms and molecular mechanisms is many–many.

The preceding concerns only single-step reactions, but as mentioned earlier, most structural proteins are produced by long biosynthetic pathways. Alterations in early steps of these reactions can affect all subsequent steps. To make matters worse, the early stages of biosynthetic pathways tend to be interlocking. The same substrate can be used in different pathways, and the same product can be produced by alternative pathways. Initially, one possible mechanism for epistasis is that the epistatic gene functions in an earlier step of a biosynthetic pathway than does its primary gene. One gene does not "mask" another. Rather the epistatic gene produces a necessary precursor in the pathway leading to the step controlled by the primary gene. Thus it is easy to see why mutations resulting in albinism are so common and why there is seldom a single gene for albinism. Biosynthetic pathways leading to the production of a pigment generally proceed for many steps before the first pigment is produced. If this pathway is blocked at any one of these numerous steps, no pigment is produced and the individual is albino. Any of the loci controlling any of the early steps is potentially albinic.

When one combines all of the complexities mentioned thus far with the requirement that the relations between the original Mendelian predicate terms be retained, one begins to appreciate the scope of the task confronting a serious proponent of reduction in genetics. Not only must he list all the mechanisms that eventuate in recessive epistasis as well as those that eventuate in dominant epistasis, but also these two lists of molecular mechanisms must be symmetrical in the same way traits that are simply dominant or recessive are symmetrical. According to the current analysis of reduction supplied by the logical empiricists and their contemporary descendants, a certain amount of reshuffling and reclassification of the phenomena of the reduced theory is to be expected, but in this instance it seems as if the vast scaffolding of Mendelian genetics must either be ignored or dismantled and reassembled in a drastically different form. As M. W. Strickberger has observed:

In the early literature the phenomena of gene action and interaction were little known. To many geneticists, the activity of any particular gene appeared to produce phenotypic effects independently of all or most genes. The discovery of instances in which the effect of one gene changed or modified that of another gene was therefore always of special interest. As more examples of gene interaction were found, a series of terms arose which were used to describe some of the different kinds of interaction. Some of these terms, such as *complementary genes, epistasis,* and others we have mentioned, are still in use today because they help us visualize the phenotypic behavior of particular genes. Actually, of course, the phenotypes we observe are the result of more complex processes than we first imagined. Epistasis, for example, is not to be taken literally as one gene "hiding" another, but may be a consequence of increased or decreased enzyme activity, changes in acidity that affect the appearance of color, or various other complex events. Whatever term we use for a gene at the gross phenotypic level, the particular developmental basis for its activity must be studied carefully and unraveled.[13]

At this point, one might object that perhaps the same molecular mechanism can result in different phenotypic effects, but that is because relevant factors are being left out—the temperature, hydrogen-ion concentration, and so on. Once all these relevant factors have been included, the one–many relation from the molecular to the phenotypic level will have been converted to a one–one relation. The conviction that this always can be done is actually a covert restatement of the principle of deterministic causality. The same cause always produces the same effect. If the effects are different, then the causes must be different. But one should notice how much the notion of molecular mechanism has been expanded. We are no longer correlating Mendelian predicate terms with molecular mechanisms but with the entire molecular milieu. One possible difference between those biologists who consider themselves reductionists and those who consider themselves organicists might well hinge on just this distinction. Reductionists think that a specification of just molecular structures and mechanisms is adequate to explain all hereditary phenomena. If so, reduction would be a straightforward, if somewhat complicated exercise. Organicists, on the other hand, emphasize that reference to the entire molecular milieu, including the environment, may well be necessary. If so, the inherent feasibility of the reductionists' program is called into question.

REDUCTION IN GENETICS— REDUCTION OR REPLACEMENT?

We began this chapter by describing a view of theory reduction popularized by the logical empiricists and the changes that have been made in this view in response to subsequent objections. We have all but ignored the more fundamental and "philosophical" objections that have been raised against the reductionist program as such. A discussion of these objections will be postponed until the last chapter. Rather we have criticized the logical empiricist analysis of reduction while

[13] *Genetics* (New York: Macmillan and Co., 1968), p. 202.

staying within that tradition. According to the logical empiricists, reduction is a formal relation between reconstructed theories. When the appropriate reduction functions have been formulated and corrections made, the reduced theory can be derived from the reducing theory. Like explanation, reduction is analyzed in terms of inference. In this chapter we have tried to show how difficult such a derivation actually is in the case of genetics. Given Mendelian genetics as it now stands, the relation between Mendelian and molecular terms is many–many. To convert these many–many relations into the necessary one–one or many–one relations leading from molecular to Mendelian terms, Mendelian genetics must be modified extensively. Two problems then arise—the justification for terming these modifications "corrections" and the transition from Mendelian to molecular genetics "reduction" rather than "replacement."

Transmission geneticists claim no interest in the variety of biochemical mechanisms that eventuate in the characters whose transmission they follow, unless these differences are reflected in different Mendelian ratios. If not, this additional information about biochemical pathways is of no use to them. However, if reduction functions are to be established that associate the terms of transmission and molecular genetics, the important biochemical differences must be read back into transmission genetics. If the same molecular structure is produced in two different ways, this difference must be reflected in the relevant reduction functions and Mendelian genetics changed accordingly. It seems strange to call these changes "corrections" because they are not required by any incorrectness or inadequacy in the way transmission genetics performs its traditional tasks. When the principles and techniques of transmission genetics are applicable and are not pushed past their power of resolution, they are completely adequate. Additional changes are required *only* because one wishes to deduce Mendelian genetics from molecular genetics.

The need for modifying the original theory being reduced is now explicitly recognized in the logical empiricist analysis of reduction, but how much can a theory be modified and still be termed the "same" theory? With enough ingenuity the phlogiston theory of heat could be reformulated so that it could be derived from modern statistical mechanics, but who would be impressed with such a "reduction"? The answer provided by the logical empiricists is that the original and corrected versions of the theory must remain strongly analogous. In the case of genetics, the two requirements seem incompatible. The amount of reconstruction necessary to permit the deduction of transmission genetics from molecular genetics would seem to preclude the existence of any strong analogy between classical Mendelian genetics and reconstructed transmission genetics. Because the notion of strong analogy was introduced to distinguish between reduction and replacement, and because this notion itself has received inadequate analysis, little can be done to resolve this conflict. (A closely analogous story can be found in the literature concerning the reduction of mind to brain states, but in this case neither the psychological nor the physiological theories are developed sufficiently to compare them with any degree of accuracy.)

If the preceding argument is cogent, then the logical empiricists are confronted by a dilemma. If the logical empiricist analysis of reduction is correct, then Mendelian genetics cannot be reduced to molecular genetics. The long-awaited reduction of a biological theory to physics and chemistry turns out not to be a case of "reduction" after all, but an example of replacement. But given our pre-analytic intuitions about reduction, it *is* a case of reduction, a paradigm case. However, if biologists are in the process of reducing Mendelian genetics to molecular genetics and the logical empiricist analysis or reduction is inapplicable to this case, the logical empiricist analysis of reduction is inadequate. The crucial issue here is whether reduction is something scientists do or an after-the-fact exercise performed on the products of these efforts. The crucial observation is that no geneticists to my knowledge are attempting to derive the principles of transmission genetics from those of molecular genetics. But according to the logical empiricist analysis of reduction, this is precisely what they should be doing.

If my analysis in this chapter is correct, then the conclusion seems inescapable that the logical empiricist analysis of reduction is not very instructive in the case of genetics. For my own part, I found that it hindered rather than facilitated understanding the relationship between Mendelian and molecular genetics. At the very least, the formal relations between these two theories are so complex and the reconstructions necessary for any detailed derivations so massive that the effort to reduce Mendelian genetics to molecular genetics hardly seems worth the effort. No one is currently engaged in producing such a derivation, and there seems no good reason to do so. The substantiation of the analysis presented in this chapter is not an easy matter, because it concerns what cannot be done. Disproof of this thesis, however, though difficult, can be decisive. All one has to do is to provide the reduction functions necessary for the derivation of Mendelian genetics from molecular genetics, carry out the derivation, and then present an analysis of "strong analogy" such that the corrected version of Mendelian genetics is strongly analogous to the uncorrected version.[14]

[14] For additional commentary, see M. Ruse, "Reduction, Replacement and Molecular Biology," *Dialectica,* 25 (1971), 39–72, M. Simon (1971), and D. Hull, "Reduction in genetics—Biology or Philosophy?" *Philosophy of Science,* 39 (1972), 491–99.

CHAPTER TWO

The Structure of Evolutionary Theory

SCIENTIFIC THEORIES, LAWS, AND DEFINITIONS

The preceding chapter concerned the "reduction" of Mendelian to molecular genetics. In this chapter we turn our attention to a second great biological theory, the theory of evolution. Just as it is not easy to draw sharp, uncontroversial lines between classical Mendelian genetics, modern transmission genetics, and molecular genetics, it is not easy to distinguish various stages and versions of evolutionary theory. For the purposes of this book, I have divided the development of evolutionary theory into three stages—classical, genetical, and synthetic. The classical version of evolutionary theory will be roughly that set out by Charles Darwin during his lifetime. The genetical theory of evolution will be the formulations produced by geneticists and mathematicians in the first third of this century. As might be expected, the genetical theory of evolution is mathematical in notation and deals with evolution in terms of gene and genotype frequencies. This single-minded concern in the genetical theory of evolution with genes has been intentionally emphasized (perhaps exaggerated) in the following pages in order to draw attention to the task confronting contemporary evolutionists—the need to synthesize Darwin's phenotypic version of evolutionary theory with the genetical version. Hence, this attempt is termed the synthetic theory of evolution. Others might wish to make distinctions other than those drawn in this chapter, but for the purposes of this volume, these are the distinctions that proved to be the most instructive.

From its inception, philosophers and scientists have not been satisfied with evolutionary theory as a scientific theory. Numerous objections have been raised to it. It is false. It is unfalsifiable. One of its basic principles, the claim that the fitter organisms tend to survive to reproduce themselves more frequently than those organisms that are less fit, is tautological. The organization of evolutionary theory is too loose to permit any judgments about its logical form. It entails inevitable progress. It precludes any estimations of higher or lower. It does not provide the necessary basis for predictions about the future development of particular lineages but can be used to explain these events once they have occurred. Most of these objections to evolutionary theory rest on ignorance or the failure to make some fairly simple distinctions. Several, however, concern fundamental issues in the philosophy of science.

The purpose of this chapter will be to investigate the logical structure of evolutionary theory and to compare it with the traditional analyses of scientific theories current in the philosophical literature. Is there something peculiar about the form of evolutionary theory? Perhaps the necessary evidence will never be available, but if it were, would the structure of evolutionary theory preclude prediction, verification, and the like? These questions are not merely academic, because the future of the human race may ultimately hinge upon their solution. Ecology is the study of short-term evolution. If all evolutionary development, including man's, is inherently unpredictable, then all the furor over ecology, the population explosion, and eugenics is empty polemics. If there is no way to predict the future development of any species, then there is no way to predict the future development of *Homo sapiens*. We may be on our way to rapid speciation, gradual evolution, instant extinction; who knows? Although considerable space will be devoted in this chapter to content, it should be kept in mind that our primary concern is the *form* of evolutionary theory and ways in which it might differ from traditional analyses of scientific theories.

The currently popular paradigm of a scientific theory is that of a deductively related set of laws. Laws in turn must be universal statements relating classes of empirical entities and processes, statements of the form "all swans are white" and "albinic mice always breed true." Just as scientific laws must be universal in form, the class names which occur in these laws must be defined in terms of sets of traits which are severally necessary and jointly sufficient for membership. That is, given the definition of a particular class, each member of that class must have all the traits listed in its definition and any individual which has all those traits is necessarily a member of that class. For example, all triangles are three-sided closed figures, and anything which is a three-sided closed figure is a triangle.

However, certain authorities are willing to accept something less than the deductive ideal in the organization of scientific theories. The laws in such theories might also be less than universal in form. Perhaps statistical laws, possibly even approximations, trend, and tendency statements might count as genuine scientific laws. Certain authorities are also willing to countenance definitions which depart from the traditional ideal of sets of conditions which are severally necessary and jointly sufficient. Perhaps definitions in terms of properties which occur together quite frequently but without the universality required by the traditional notion of definition might be good enough. For example, the names of biological species as temporally extended lineages can be defined only as cluster concepts. That is, no property which can be used to distinguish the species under investigation from all other species is universally distributed among the members of the species and vice versa. At best, membership in the species is determined by an individual possessing enough of the most important characters listed in its definition (see Beckner, 1959).

Thus, at one end of the spectrum are deductively organized theories that contain nothing but laws that are universal in form and terms that are defined in the traditional manner; at the other end are inductively organized theories that contain less than universal statements and terms that can be defined by properties which vary only statistically. The characteristics of the first type of theory and definition have been worked out in great detail by logicians and philosophers. The precise nature of the second kind is still highly controversial. The question which will concern us in this chapter is whether evolutionary theory can be made to approach the deductive ideal with respect to the form of both its laws and its definitions—and if it cannot, whether this failure reflects inadequacies in evolutionary theory or in the currently accepted paradigms of scientific theories and definitions.

Another traditional requirement of scientific theories is that the laws which make them up must be unrestricted temporally and spatially. This does not mean that spatial and temporal variables have to be excluded from scientific laws—to the contrary, they play important roles in theories throughout science. Rather it means that scientific law cannot be limited *necessarily* to a particular time span or place in the universe. A scientific law can be limited to planets while they are part of effectively closed star systems. There might even be only a single planet in the universe revolving around a single star. But a scientific law cannot deal just with Mars on June 15, 1935. It can deal with an event that happens to be one of a kind—just so long as it is of a kind. It cannot deal with *necessarily* unique events. A unique event is one that happens to be one of a kind. A necessarily unique event is defined in terms that preclude any other instance of this event. The usual way of defining an event so that it is necessarily unique is by specifying particular spatiotemporal coordinates for it. No two objects can occupy the same place at the same time.

By and large philosophers define the notion of "individual" in terms of spatiotemporal extension and continuity, and their definition coincides quite closely with the ordinary meaning of the term. The things which we commonly call individuals are extended continuously in space and time and have reasonably sharp boundaries. Since individuals are by definition spatiotemporal entities, it follows *a fortiori* that scientific laws cannot refer to specific individuals, only to classes of such individuals. The reader should be warned, however, that "individual" is not being used here in any absolute sense. Ordinarily we as human beings are used to treating certain complexes as individuals. Because multicellular organisms tend to be both small enough and large enough for us to observe with our unaided senses and because they tend to remain intact long enough relative to our ability to perceive them, we treat them as paradigm individuals. Scientists need not. A scientist could devise a theory in which individual organisms function as classes, perhaps classes of cells. Another scientist might devise a theory in which individual organisms are parts, perhaps parts of biological species. A pervasive problem throughout the history of evolutionary theory is the discovery of evolutionary units which function adequately in evolutionary laws. Biologists can be found saying that individuals at nearly every level of analysis are *the* units of selection—genes, parts of chromosomes, whole chromosomes, genotypes, organisms, Mendelian populations, biological species, and so on.[1] But the point I wish to make here is that whatever entities a scientist might pick to be the individuals in his theory, the names of those individuals cannot function in the laws of that theory—only the names of classes of those individuals can do that. For example, if particular chromosomes are the individuals in a certain version of evolutionary theory, then a law in that theory could refer to the class of sex chromosomes but not to a particular sex chromosome in a particular cell.

In the face of such apparent disagreement among scientists, one might be tempted to resort to some ontologically ultimate individuals like sense data and bare particulars, whose status is independent of any scientific theory. But the history of such individuals does not make this alternative too attractive. Instead it seems necessary to recognize various levels of individuality, depending on the scientific laws under consideration. Relative to one set of laws, a certain complex may be treated as an individual. Relative to another set of laws, it may be treated as a class. Because all these "individuals" are part

[1] For the appropriate references, see Mary B. Williams, "The Logical Status of Natural Selection and Other Evolutionary Controversies: Resolution by an Axiomatization," *Philosophy of Science Symposium* (Dordrecht: D. Reidel, 1973).

of one and the same reality, it is tempting to assume that there *must* be one perspective from which all the various levels of analysis can be integrated into a single system and one all-encompassing theory formulated which is capable of dealing with it. This is the notion of the unity of science, a notion whose reputation has been unnecessarily damaged by premature attempts at such an integration and a dogmatic insistence that it is the only aim of science. Scientific theories do exist at various levels of analysis; they should be evaluated on their own merits rather than just in terms of the extent to which they can be reduced to the next lowest level.

One final comment must be made about the nature of scientific laws before we proceed to an analysis of evolutionary theory. The currently accepted paradigm of a scientific law is a process law, a generalization that permits one to infer the state of a system with respect to the values of certain variables at any time whatsoever from knowledge of the state of these variables at any one time. Process laws are possible only for systems like the solar system that are effectively closed with respect to the relevant variables. Given the laws of celestial mechanics and knowledge of the position and momentum of each of the heavenly bodies in the solar system at any one time, the positions and momenta of these bodies can be inferred for any other time, earlier or later, with a high degree of accuracy—as long as sufficient closure is maintained. Because problems involving so many bodies are prohibitively complex, rarely are more than two bodies treated at a time, with a consequent loss of accuracy. Furthermore, if an extremely large comet were to pass through our solar system, it would disrupt it so that all previous calculations for future dates would be erroneous. The laws of celestial mechanics would not have been shown to be false; rather, closure would have been breached.

Time and again it has been observed that rarely are biological systems sufficiently closed for long periods of time; biosynthetic pathways, individual organisms, breeding populations, ecosystems, and entire species are partially and often necessarily open systems. But there are strategies that can be employed to circumvent this difficulty. Perhaps process laws can be produced for the behavior of biological systems during intermittent periods. Just as physical laws can be used to infer the positions of the heavenly bodies in our solar system only when it is effectively closed, biological laws can be used to infer values for the relevant variables of a biological system only so long as and whenever it is sufficiently closed. In such cases, a biological law would not necessarily apply to a species throughout its existence but only in special circumstances; for example, when some of its members enter a previously unoccupied ecological niche or when one of its populations becomes reproductively isolated from the main body of the species. Similarly, a physiological law might not apply to an organism throughout its existence but only during a specified stage in its development.

DARWIN'S ARGUMENT AND THE STRUCTURE OF EVOLUTIONARY THEORY

In order to talk sensibly about evolutionary theory, it is necessary to stress the seemingly obvious differences between evolutionary processes and phylogeny and between evolutionary theory and phylogenetic descriptions. Evolutionary processes are the mechanisms by which species evolve; phylogeny is the product. Evolutionary theory characterizes the process; phylogenetic descriptions, the product. As Hempel (1965: 370) admonishes his readers, we must "distinguish what may be called the *story* of evolution from the *theory* of the underlying mechanisms of mutation and natural selection." The claim that mammals arose in the Jurassic Period from ancestral reptiles is part of the story of evolution, a phylogenetic description, and not part of evolutionary theory. On the other hand, the claim that speciation can take place only in the presence of appropriate isolating mechanisms is part of one version of evolutionary theory. Phylogenetic descriptions are historical narratives; evolutionary theory is not.

One must also distinguish between Darwin's argument for the existence of evolution and a full blown theory of evolution. The outline of Darwin's argument for the existence of evolution is easy to follow. Organisms tend to produce young in such quantities that if they all lived to reproduce themselves in the same plentitude and their offspring did the same indefinitely, the earth would rapidly be overpopulated even by a single species. Clearly this does not happen. Some species increase in numbers seasonally, only to decrease later. Others increase during certain geological periods only to become extinct later. Fluctuations do occur in the number of organisms that exist at any one time in the history of the earth, but there is nothing like the rapid expansion that one would expect on the basis of their reproductive potentiality. Hence, Darwin reached the rather obvious conclusion that a high percentage of the organisms produced die before they in turn are able to reproduce.

Two possibilities present themselves at this point. Either the destruction of all these organisms could occur entirely at random or it could result at least in part from the relation between each individual organism and its environment—the environment including, of course, other organisms of its own and other species. Certainly many organisms die in a haphazard manner unrelated to any differences between them and other organisms. For example, few if any organisms could survive immersion in lava, regardless of any specific characteristics they might or might not have. In some cases however, it seems likely that there is a differential survival of organisms that is a function of the relation between them and their environment; they survive because in the main they are better adapted to their environment than other organisms are to theirs. In the struggle for existence, the fitter organisms tend to survive more frequently than the less fit.

If all members of a species were identical and were exposed to the same environments, which members of a species survived to reproduce and which did not would be a matter of chance. Fitness could be invoked to explain only why one species persisted and another became extinct. The existence of intraspecific variation is necessary if speciation is to occur. Certain characteristics that organisms possess are heritable; regardless of the mechanisms involved, they tend to be passed on from generation to generation. If some of the variations prevalent among the organisms that comprise a species are heritable and continue to vary in the same direction indefinitely, then through the succession of generations one would expect organisms to become better and better adapted to their environments. If the environment of a species were identical throughout its range and remained the same, a species would reach a point of optimum adaptiveness and oscillate around this point. But if environmental variation occurred in the range of this species or if the environment changed successively in time, new species might evolve.

For example, the range of a particular species might extend from an arid to a mesic region (from a dry region with occasional rain to one with a balanced supply of moisture). If this species were divided into two separate populations by some sort of barrier, it might split and evolve into two new species, since the organisms in the arid region would tend to survive because of one set of adaptive characters and those in the mesic region because of another set. In short, they would be exposed to different selection pressures. Or a single species might evolve intact progressively in time adapting to a progressively changing environment. For example, it might adapt to a mesic environment as the conditions throughout its range gradually changed from arid to mesic. The former process is termed "speciation"; the latter "phyletic evolution."

SALTATIVE EVOLUTION AND ESSENTIALISM

We tend to think of the initiators of great scientific revolutions as breaking completely with the past—something they never do. Just as Copernicus retained Ptolemaic epicycles, Darwin retained vestiges of earlier biological theories in his work. Although he maintained throughout his life that natural selection was the chief directive force in evolution, he accepted the minor influence of use and disuse and the direct effect of the environment. His theory of pangenesis included mechanisms by which both types of effects might occur. Through the years Darwin modified his theory somewhat, but two tenets which he steadfastly refused to change were that evolution occurs gradually and is in no significant sense directed toward some goal or goals. Biologists rapidly admitted that species evolved, but they raised loud objections, both to the mechanisms proposed by Darwin and to the preceding two tenets. The objections of St. George Jackson Mivart (1827–

1900) were among the most influential and were in many ways typical. Mivart argued that species evolved saltatively in large, abrupt steps guided by some unknown internal innate force. Species evolved like a self-propelled, faceted spheroid, tumbling from one facet to another. The reason so many biologists in Darwin's day preferred to think of evolution as being directed will be discussed in the chapter on teleology. Their preference for saltative evolution also had a metaphysical foundation; it would allow the retention of the essentialist notion of definition and hence natural kinds.

Philosophers from Socrates on have maintained almost unanimously that class names must be defined in terms of properties that are severally necessary and jointly sufficient for membership in the class; that is, each of the properties taken separately is necessary for membership and the whole set taken together is sufficient. Incorporated into the definition of any natural kind will be the name of a higher-level natural kind in which it is included. For example, a bachelor is a man who has never married. "Man," in turn, is defined as an adult, male human being. "Human being" in turn is defined as an animal possessing an opposable thumb, plantigrade foot, highly developed brain, and so on. This traditional mode of definition is admirably suited for a world made up of nested hierarchies of discrete natural kinds; it is not, however, very appropriate for characterizing gradually evolving species. If Darwin's theory is accepted, then neither species nor the properties used to define them are discrete entities. Lungs gradually develop in certain lines and disappear in others; at no one point does a structure become a lung or cease to be a lung. Similarly, species evolve as these structures gradually change; at no one point does one species become another. Evolving lineages can be divided into species, but the names of these species can be defined in the traditional manner only by highly artificial devices, typically by counting a single character as distinguishing between species. Attempts to define "human being" in terms of a single essential characteristic are typical. The traditional mode of definition is salvaged but at the price of the significance of the resulting concept. If contemporary taxonomists are in agreement about anything, it is that species of plants and animals cannot be distinguished by means of a single character without sacrificing the purposes of biological classification. In saltative evolution, numerous characteristics change abruptly in the space of a single generation. Many of Darwin's contemporaries were persuaded to accept evolution, but only if it were saltative—their preference merely reflected their predisposition for the essentialist mode of definition. Perhaps species were not eternal, but at least they might be discrete (see Hull, 1973).

Both before and after the rediscovery of Mendel's laws, biologists argued over the nature of the variations operative in evolution. Were they large and abrupt as Hugo de Vries and later Richard Goldschmidt claimed? If so, then evolution was saltative. Were they small and abrupt as Francis Galton, William Bateson, and T. H. Morgan maintained? If so, was evolution gradual or saltative? One of the confusions that plagued evolutionary theory after the advent of modern genetics was the equating of the size of the mutation with its abruptness. It is easy to see why de Vries found himself in opposition to Darwin, but Galton, Bateson, and Morgan also thought that their views conflicted with those of Darwin. Darwin did not claim that the variations that provided the raw material for evolution were "continuous." This was Bateson's term. Rather he said that they were "slight" or "imperceptible" in contrast to those that were "strongly marked." In addition, a whole series of them did not occur at once. Hence, evolution was gradual.

Robert Olby (1966: 82), in his otherwise excellent account of blending and nonblending heredity, asks why Darwin did not accept the principle of evolution by many small but definite steps. To the extent that the relevant distinctions had been made at the time, he did. Darwin was concerned with excluding evolution by large, abrupt phenotypic changes. It did not matter whether these changes were caused by a single massive change in the genotype or by numerous, simultaneous small ones. Darwin's intuitions on the matter were basically correct. The only genetic mechanism that has been discovered that produces the saltative evolution Darwin argued against is polyploidy; but in the vast majority of cases, evolution is gradual, not saltative. (See also Vorzimmer, 1971 and Provine, 1971.) Thus, if natural kinds in the classic sense are necessary for scientific laws, the names of species can never function in natural laws. Hence, the statement "all swans are white," even if true, is not even a candidate for a law of nature. Only if the notion of definition is expanded to include definitions in terms of statistically covarying characteristics can the names of species and higher taxa appear in natural laws.

PANGENESIS AND BLENDING INHERITANCE

To accommodate the wide range of hereditary phenomena which Darwin believed to exist and to provide the laws of inheritance which evolutionary theory required, Darwin formulated his theory of pangenesis. According to this theory, all parts of the body produce gemmules (or pangenes) characteristic of that part of the body. These gemmules congregate in the gonads in varying proportions to be passed on in reproduction. The characteristics of the resulting offspring depend both on the quantity and the quality of the gemmules received from the parents. If an individual receives more gemmules for brown eyes than for blue, it will have brown eyes. Sometimes gemmules are latent, being passed on in a dormant state from generation to generation until for some unknown reason they begin to proliferate to produce a reversion. Because gemmules are produced throughout the body, changes in the body might affect the related gemmules, allowing use and disuse and the environment to affect inheritance.

Commentators frequently bemoan the fact that Darwin toyed with the idea of particulate inheritance in 1857 only to forsake it for his theory of blending inheritance.

> Approaching the subject from the side which attracts me most, viz., inheritance, I have lately been inclined to speculate, very crudely and indistinctly, that propagation by true fertilization will turn out to be a sort of mixture, and not true fusion, of two distinct individuals, or rather of innumerable individuals as each parent has its parents and ancestors. I can understand on no other view the way in which crossed forms go back to so large an extent to ancestral forms. But all this, of course, is infinitely crude.[2]

Here Darwin is not anticipating Mendelian particulate inheritance but is referring to his own theory of pangenesis. Darwin's theory is "particulate" in the sense that his gemmules are particles. It is "blending" both because the gemmules themselves on occasion fuse and because the number of gemmules which cooperate to produce a character is large and variable. Whether hybrids are intermediate in character between the two parental forms or whether one form is "preponent" over the other, depends on the "number, vigour, or affinity of its gemmules." Because each cell throws off so many gemmules of varying quality and because a varying percentage of each kind of gemmule is transmitted to any one zygote in fertilization, almost anything can happen. But on an average, the number of any one kind of gemmule is reduced by half at each mating. This is the sense in which pangenesis is a "blending" theory of inheritance—contrary to the analysis of Darwin's ideas on heredity provided by Fisher in 1930 and by Vorzimmer in 1971.

Darwin's theory of pangenesis was never very popular and fell into disrepute, though it was revived in a modified form by de Vries in 1889. The problem with this theory was that it was indefinitely expansible. Whatever happened could be accounted for *ex post facto,* yet no precise prediction could be made. There was no reason to expect crosses between blue-eyed and brown-eyed people to produce children of one eye color rather than another, let alone in any definite proportion. One would expect acquired characters to be heritable because gemmules were produced throughout the body, but if not, this fact would not necessarily falsify pangenesis. Gemmules could be modified by the environment; they just did not happen to be. Gemmules did not need to be produced in any fixed ratio, but if they were, fine. This did not count against pangenesis either. About the only thing that could refute pangenesis was the discovery that the hereditary material that exists in somatic cells in no way travels to the gonads to be passed on in reproduction.

[2] Francis Darwin, ed., *More Letters of Charles Darwin,* I (London: John Murray, 1903), 103.

Proof for this claim could not be made conclusive until August Weissman clearly distinguished between germ cells and somatic cells.

SINGLE VARIATIONS AND INDIVIDUAL DIFFERENCES

The distinction important for Darwin was between single variations and individual differences. A single variation was the occurrence of a single unusual individual. Darwin typically thought of such single variations as being marked—what are commonly called "sports." Individual differences were the slight differences that occurred quite frequently among the members of a species. Darwin believed that individual differences were the source of evolutionary change, not single variations. A paper by Fleeming Jenkin (1867, in Hull, 1973) convinced him that he was right on this score but raised serious doubts as to the adequacy of evolution by natural selection to produce new species on the supposition of blending inheritance.

Jenkin argued that on Darwin's principles of inheritance, a new character could not become established in a population no matter how drastic the selection. For example, a white man might be introduced into a population of one hundred blacks. If he killed all the black males and then mated with all the black females, the next generation would all be medium brown. If he were then to kill all of his sons and his former wives and then mate with his daughters, the next generation would be 75 percent white, and so on until the man died. The shade of skin then prevalent would remain unaltered until the tribe received another visitor, either darker or lighter in skin color. Thus, on the supposition of blending inheritance, a white man could not turn a nation of blacks white.

Jenkin's conclusion reaffirmed Darwin's belief that the numerous slight variations observed at all times among the members of a species were the source of evolutionary change. For example, in a particular tribe not all the members will be uniformly the same color. On a scale of 10, the darkest member of the tribe might register 10, the lightest 8, and the mean shade for the entire tribe be 9. Assuming for the moment that lighter skin is more advantageous, more of the lighter-skinned individuals would survive to reproduce than those with darker skin. In the next generation, the mean shade might be 8.5. In the original population, color varied one point in each direction from the mean. If these extremes remained the same from generation to generation, the darkest members of the tribe would register 9.5 on the scale and the lightest 7.5. Thus, even on the supposition of blending inheritance, evolution could occur if the characters continued to vary "in all directions" from one generation to the next.

Jenkin seized upon the weakest assumption in Darwin's argument. Even though the mean might be shifted by natural selection, it does not follow that limits of variation will also shift. Jenkin supposed on the contrary that a sphere of variation existed for each species. Initially selection could make great progress in shifting the mean value for a character, but as the limits of variation were approached, further progress became more and more difficult to bring about. Furthermore, when the selection pressure for this particular extreme was removed, the population would return of its own accord to its original distribution. To bolster his conviction, Jenkin cited the experience of horticulturists and animal breeders. Darwin had to admit that Jenkin's argument was very forceful, but he still retained his conviction that these hypothetical limits of variation could be transgressed. The phenomena explained by evolutionary theory were too numerous and too diverse for it to be wholly wrong.

Darwin's mistake, of course, was his dependence on a theory of inheritance that results in most characters blending upon transmission, his theory of pangenesis. He was aware that in heterogeneous matings some characters disappeared completely only to reappear unmodified in later generations, but he viewed such occurrences as reversions to type, a phenomenon that could hardly function as a foundation for evolution!

Jenkin, unwilling to leave Darwin any avenue of escape, entertained the possibility of an organism's being able to pass on a character indefinitely and undiluted to its descendants. On such a supposition, evolution could occur at a rapid rate. He wrote, "But this theory of the origin of species is surely not the Darwinian theory; it simply amounts to the hypothesis that, from time to time, an animal is born differing appreciably from its progenitors, and possessing the power of transmitting the difference to its descendants. What is this but stating that, from time to time, a new species is created?"

In the above quotation Jenkin exhibited a blindness and prejudice which was later to be equally characteristic of the founders of Mendelian genetics. A change in a single character, no matter how abrupt, is hardly sufficient for the recognition of a new species whether in Darwin's day or at the turn of the century, nor could this process be called "creation" in the supernatural sense that the term carried then. The importance of evolutionary theory can be estimated by the fact that all the early work in Mendelian genetics was performed with one eye to its implications for evolution. The facility with which many of these investigators claimed to have refuted evolutionary theory attests to how poorly they understood the theory and how biased they were against it.

In the dispute over the nature of the hereditary component in evolution, everyone was partly right and partly wrong. Darwin was correct in thinking that the variations that are most significant in evolution are slight. He was mistaken in thinking that the variations commonly observed among the members of a species can all equally serve as raw material for evolution. He did not know that most individual differences are due to recombination and the variable effects of the environment. Only a small percentage of these slight differences are actually "mutations" in the sense of substantive changes in the hereditary material. The rarity of mutations is compensated for by particulate, nonblending heredity. Observable characters may appear to blend but genes do not, and genes come in pairs, not variable clusters.

THE GENETICAL THEORY OF EVOLUTION

In spite of the initial flap about the incompatibility of Mendelian genetics and evolutionary theory, biologists soon came to see that these two theories were fully consistent (see Provine, 1971). Darwin had formulated his theory of evolution assuming that there must be some mechanism for the production and maintenance of the type of variation it required. Mendelian genetics was precisely that theory. One had every right to expect that a combining of Mendelian genetics with Darwin's theory of evolution would have resulted in a more powerful evolutionary theory. It was with this end in mind that R. A. Fisher, J. B. S. Haldane, and Sewall Wright set out to construct what has come to be known variously as the mathematical, the Neo-Darwinian, or the genetical theory of evolution. Although the work of these three men differed in some important respects, all of their formulations had two things in common—they were mathematical in notation and they concerned the distribution of genes in successive generations.

The initial role of mathematically inclined biologists in the development of evolutionary theory was to show that certain commonsense notions about hereditary transmission and the power of natural selection were mistaken. For example, some biologists wondered why all characteristics that existed in dominant and recessive states did not come to exist in 3-to-1 ratios in all natural populations. After all, if equal numbers of organisms heterozygous for a trait are crossed, the resulting offspring exhibit a 3-to-1 ratio, and in all succeeding generations this ratio will be maintained if mating is random and if significant selection pressures are absent. In 1908, G. H. Hardy and C. Weinberg independently pointed out how faulty these mathematical intuitions were. In cases of mutation, a single instance of a new allele is introduced into a population, not an equal number of heterozygotes. Regardless of whether this mutation is dominant or recessive, it will be maintained at a very low percentage of the whole population, provided it is selectively neutral. Hardy and Weinberg argued that in a large population in which males and females mated at random with respect to the gene in question and produced equal numbers of equally fertile offspring, *ceteris paribus* one should expect the maintenance of roughly the *same* ratios of the various homozygotes and heterozygotes. If we let p equal the percentage of gametes carrying one allele and q the percentage of gametes carrying the other allele, then the Hardy-Weinberg law can be expressed simply by the binomial equation,

$$p^2 + 2pq + q^2 = 1,$$

where p^2 represents the frequency of one homozygote, $2pq$ represents the frequency of the heterozygotes and q^2 represents the frequency of the other homozygote. The genetical theory of evolution might be mathematical, but not very.

The reason for the importance of the Hardy-Weinberg law was not its mathematical notation, but the fact that it provided a simple standard of comparison for actual natural populations. If the gene frequencies predicted by the Hardy-Weinberg law did not obtain from generation to generation, then everything must not have been equal. Perhaps the matings were not random with respect to the new gene. Perhaps additional new genes were being introduced into the population, or some selection pressure was affecting the population. In any case, some agency must have been at work. The simplicity of the Hardy-Weinberg law, however, was purchased at a price. It contained explicit reference only to gene and genotype frequencies. The intricate relation between individual genes and the resulting phenotype and between individual organisms and their environments was ignored in the mathematical formulae that formed the genetical theory of evolution. The only prediction possible from the Hardy-Weinberg law is that the gene ratios under investigation will remain constant and, if they do not, something must be causing the change. There is no way, for example, to predict that an observed increase in the frequency of p^2 will continue, stop, or reverse. To make such predictions one must go outside the mathematically characterized scope of the genetical theory. The Hardy-Weinberg law as a formula merely indicates how the percentages of the possible combinations of the two elements can be related. The elements could be two pennies. It is the boundary conditions that supply the empirical content to the law.

THE STATISTICAL NATURE OF EVOLUTIONARY THEORY AND ITS SOURCES

One of the prerequisites of a scientific theory is that it must permit inferences about natural phenomena. These inferences can be deductive or inductive. They can be about past, present, or future events. In the typical case, they are deductive inferences about the future; i.e., deductive predictions. Evolutionary theory might permit several types of predictions: predictions about a particular organism (like Gargantua), about a particular species (like *Phthirius pubus*), about some grouping of organisms not now commonly recognized by biologists (several will be mentioned in the following pages), or about various possible patterns of evolution or kinds of species (such as species undergoing adaptive radiations). In the past, biologists have concentrated on predictions about particular species or populations. We will argue in this chapter and the next that this emphasis has been misplaced. Species as they are currently defined may not be the most significant evolutionary units. Of greater significance is the conclusion that any putative laws of evolution will make no direct reference to any particular class of organisms, regardless of how it may be defined. Instead they will refer to kinds of species, populations, or some other unit.

The paradigm of a scientific theory in Darwin's day was Newton's mechanics. It was deductively organized and permitted accurate predictions about individual cases. For example, given the appropriate data and the laws of celestial mechanics, one could infer the position of a particular planet, such as Mars, with a high degree of precision. On occasion, biologists in Darwin's day and today are in a position to make precise predictions about individual organisms, but not on the basis of evolutionary theory. As the philosopher C. S. Peirce recognized quite early, the form of evolutionary theory differed markedly in this respect from the Newtonian paradigm. Given the relevant data and evolutionary theory, no reasonable inferences were possible concerning the fate of a particular organism. But Peirce also suggested that on a different paradigm, evolutionary theory did not look so bad. He argued that, like the theory of gases that was gaining acceptance at the time, evolutionary theory was statistical in form and permitted inferences only about what happens to large ensembles of individuals in the long run.

> The Darwinian controversy is, in large part, a question of logic. Mr. Darwin proposed to apply the statistical method to biology. The same thing has been done in a widely different branch of science, the theory of gases. Though unable to say what the movements of any particular molecule of gas would be on a certain hypothesis regarding the constitution of this class of bodies, Clausius and Maxwell were yet able, by the application of the doctrine of probabilities, to predict that in the long run such and such a proportion of the molecules would, under given circumstances, acquire such and such velocities; that there would take place, every second, such and such a number of collisions, etc.; and from these propositions they were able to deduce certain properties of gases, especially in regard to their heat-relations. In like manner, Darwin, while unable to say what the operation of variation and natural selection in every individual case will be, demonstrates that in the long run they will adapt animals to their circumstances. Whether or not existing animal forms are due to such action, or what position the theory ought to take, forms the subject of a discussion in which questions of fact and questions of logic are curiously interlaced.[3]

If all a biologist were interested in doing was predicting that evolution would occur, evolutionary theory would be statistical, but it would be statistical in a sense that would pose no special problems for the philosophy of science. Although the behavior of each individual molecule in an enclosed gas cannot be predicted on the basis of the gas laws, the behavior of large

3 Peirce, "The Fixation of Belief," *Popular Science Monthly,* 12 (1877), 1–15.

ensembles of such molecules can be and with a high degree of accuracy, but as Darwin himself admitted, even in cases of large ensembles of individual organisms, reasonable predictions are impossible. "When we descend to details, we can prove that no species has changed; nor can we prove that the supposed changes are beneficial, which is the groundwork of the theory. Nor can we explain why some species have changed and others not."[4] It appears as if evolutionary theory is statistical at the level of such classes as populations and species. The reasons for this apparent indeterminacy in evolution must be carefully analyzed and the manageable kinds of indeterminacy distinguished from unmanageable kinds. We will argue that reasonable inferences about particular species or populations as such cannot be made on the basis of evolutionary theory. At best, predictions can be made only about these groups to the extent that they exhibit one of the patterns of evolution thus far distinguished. We will also suggest, somewhat more tentatively, that perhaps traditional biological units like species are not the appropriate units for evolutionary theory.

The statistical nature of evolutionary development of the manageable kind stems both from the genetic component of the synthetic theory of evolution and from several features of natural selection. In meiosis, both crossing over and assortment occur at random with respect to the laws of Mendelian inheritance. At fertilization, which sperm unites with which egg is again largely a matter of chance. But all of these elements could be discounted if the populations in question are large enough, just as the random movement of gas molecules can be discounted in the behavior of gases. Certainly sometimes natural populations are sufficiently large to allow prediction of the statistical distributions of genes from generation to generation. Even the randomness of mutation can frequently be ignored if the populations are large enough. Mutation rates are low, but not so low as to preclude their being counterbalanced by sufficiently large numbers. If two genetically identical populations are allowed to expand without restriction, the resulting populations will be nearly identical from a statistical point of view in spite of all the random factors mentioned.

Randomness of an unmanageable variety enters into evolution when we expand the scope of our investigation to include the variable effect of the environment on evolutionary development. The problem here is one of closure. The distinction between an individual organism and its environment is not completely unproblematic, but in most cases it can be made without confronting insuperable difficulties. The individual organism results from the successive interaction between the genome via the phenome and the environment. But the genetic makeup of an organism at least prescribes limits to the range of its possible phenotypic development. However, when we expand our interest to include such traditional units of evolution as species and popula-

[4] Francis Darwin, *The Life and Letters of Charles Darwin,* II (New York: Basic Books, 1887, reissued 1959), 210.

tions, the situation is very different. These systems are far from closed with respect to the factors affecting them. If we are to predict the course of development of a particular species, then we must be able to predict the successive environments to which the members of this species are to be exposed. Hence, we must predict both geological and cosmological changes of the earth as well as both short-term and long-term climatic changes with a high degree of precision. A river changes course and splits a single population into two, a new predator invades the territory of another species, a new mutant appears and must be accommodated or eliminated. At the time the mutant is introduced it is disadvantageous. Later the situation changes. If it were introduced then, it would become established. However, it does not appear before the environment changes once again. Ecological niches open and close without ever being occupied.

Scientists are currently not in a position to estimate values for the environmental elements that influence the development of particular populations to the extent necessary to permit much in the way of long term predictions about the future of such populations. However, even if we had the requisite data about the numerous factors that affect evolution, we currently have no way of combining this information in a meaningful way to produce detailed and precise predictions. Richard Lewontin (in Mendelsohn, 1969) characterizes the situation as follows:

> Most of evolutionary theory has the following structure: there is some general framework of laws, and we know, from observations about nature, something about the parameters that we put into the calculating engine. Thus we produce an allowable set of outcomes. We know something about mutation rates; we know about the differential rates of fertility of organisms. Given such bounds, the evolutionary biologist can then predict a possible set of outcomes. Then if he is really very clever, he can even set up a distribution function of the probabilities of the outcomes. The actual outcome is then observed, and we ask (a) whether it falls inside the set of allowable outcomes, and (b) if it does, what was the probability of that outcome.

If the outcome is one of the highly probable ones, the evolutionary biologist is happy. If it is one of the less likely ones, he tries to find the relevant extenuating circumstances. If he cannot, or if the actual outcome falls outside the set of allowable outcomes, doubt is cast on the laws, the accuracy of the original data, or else on the means by which this information was processed to produce the predictions. Evolutionists are satisfied with a set of allowable outcomes whose total likelihood is high, rather than insisting upon a prediction of a specific outcome, because the likelihood surface is usually close to being flat. There is not that much difference between the allowable outcomes.

Someone might object that the situation really is not all that much better for celestial mechanics if we want to infer future (or past) states of the solar system as such. It is no more possible to deal with all the bodies in the solar system at one time than it is to deal with all the factors affecting the evolution of a particular species. Similarly, just as evolving systems are at best only partially closed systems, the closure of the solar system could be breached at any time by a huge comet or the like. The reply to this objection consists in pointing out factual differences between the two situations. In celestial mechanics, the position of a particular planet can be predicted with great accuracy taking into account only the sun and that planet because the effects of the other planets are relatively negligible compared to the overwhelming effect of the sun. And if one wished, the effects of the other planets, could be taken into account one by one. In evolution, numerous factors enter in, and rarely is any one of them so important that the others can be ignored. In celestial mechanics there are two-body problems that are easily solved; in evolution, there are no significant two-body problems.

THE FOUNDER PRINCIPLE

Even if we ignore the problems introduced by the variable and erratic effect of the environment on evolving units and the difficulty of combining what data we do have in order to make some sort of a meaningful prediction about such units, we are still left with the fact that frequently natural populations are so small that one or more of the numerous random elements specified earlier can become significant. Ernst Mayr (1963) has argued that such small populations play a key role in evolution, especially among sexually reproducing organisms. According to him, species tend to divide into two or more species only when an ecological or geographical barrier is introduced that impedes genetic exchange. Frequently, a segregated population is very small—hence, its genetic makeup is greatly impoverished and represents a chance selection from the much more extensive range of genotypes present in the main body of the species. While the population is small and isolated, numerous factors that are negligible in large populations suddenly become very important. Genetic drift can further reduce the genetic makeup of the population by the chance elimination of certain alleles. Increased inbreeding will result in increased homozygosity, a well-known consequence of the Hardy-Weinberg law. The selective values of certain genes will be changed because of the absence of other genes with which they interacted epistatically in the parent population. And so on.

If this population is to persist, then either it cannot remain small or it cannot remain isolated for long. It must expand rapidly. It may remain isolated, become reproductively isolated, and evolve into a new species. Or it may merge again with the main body of its species, introducing a large, atypical dose of genes. In either case, this evolutionary event is of extreme importance—so important that Mayr has termed the effect that the impoverished genetic makeup of founder organisms has on the future development of any colony they might found, the founder principle. The importance of the founder principle for our purposes is that it introduces an uneliminable random element into evolutionary development, if what we are trying to predict is the uninterrupted evolution of a particular lineage. For contingent reasons, we cannot predict which organisms will survive, but one thing is certain—even if we knew which organisms would found new populations and what their genetic makeup might be, we still could not predict the future development of any colony going through such a bottleneck because of the small number of individuals involved. As L. E. Mettler and T. G. Gregg summarize the situation in their introductory text,

> Of all the innumerable genotypes possible, only a relatively few are actually formed in any generation during the life of the population. Since all populations are finite and the number of possible genotypes is unimaginably large, the great majority of the theoretical combinations are not realized simply because of chance—even some of those that are produced are later randomly eliminated accidentally.[5]

In this chapter we have identified three possible sources of the statistical nature of evolutionary theory as it is usually formulated. The first concerns lack of knowledge and is in principle eliminable, but it seems that to apply evolutionary theory to particular evolutionary units such as populations, we need a running commentary on the successive changes in the particular circumstances and boundary conditions involved so extensive that it takes on the proportions of an historical narrative (see discussion in Chapter Three). But even if we had the relevant data, we are currently not in a position to process it adequately, perhaps through the failure of biologists to discern the appropriate evolutionary units and key factors, perhaps due to the lack of the appropriate mathematical tools. But even if both of these inadequacies were remedied, we are still presented with the problem of the founder principle. The gas laws can be treated as being deterministic for enclosed gases made up of large numbers of molecules; but they apply less and less accurately as the number of molecules decreases until they do not apply at all. The problem posed by the founder principle is that laws concerning the evolutionary development of particular populations or species, if they are to apply at all, must necessarily apply to populations that are extremely small relative to the amount and variety of indeterminacy involved. Only the founder principle seems to entail that evolutionary theory, as traditionally formulated, is necessarily statistical.

THE DEDUCTIVE NATURE OF EVOLUTIONARY THEORY

In the preceding discussion, we have joined with all those biologists mentioned in assuming that the classes that must be incorporated into evolutionary laws are time slices of such evolutionary units as populations or species. Biologists have been forced to admit that if these are the relevant classes and if what one wants to predict is the future

[5] *Population Genetics and Evolution* (Engelwood Cliffs, N.J.: Prentice-Hall, Inc., 1969), p. 89.

development of these classes, then at best only probabilistic predictions are possible. But if one were to modify one's choice of the units that function in evolutionary theory and the type of prediction desired, it might be possible to organize evolutionary theory in a way compatible with the notion of a deductive scientific theory set out in the early pages of this volume. Mary B. Williams[6] has done just this in her axiomatization of evolutionary theory.

Several philosophers and biologists have argued with varying degrees of success that there is a deductive core to evolutionary theory, but by this all they mean is that from a few basic principles one can deduce that evolution will occur. More detailed deductions than this are necessary before one can claim that evolutionary theory is deductive in any significant sense. One must also produce deductive predictions that bear on the various axioms of evolutionary theory. Is evolution gradual, saltative, or a little bit of both? Can speciation occur in the face of gene exchange between populations? Can a hereditary trait be selected which regulates the population size at a level below the carrying capacity of the available resources? What percentage of extant species should be at the various possible stages of speciation? By recognizing the existence of several levels of units functioning in evolution and carefully specifying how they are related, Williams has been able to organize her version of evolutionary theory deductively and to derive a variety of predictions from it about various patterns of evolution.

As Williams sees it, the purpose of her axiomatization is "to express Darwin's theory of evolution as a deductive system in which a few fundamental principles of the theory are used as axioms from which the remainder of the principles of the theory can be deductively derived." It provides the "foundation for a precise, concise and testable statement of Darwin's theory of evolution (i.e. anagenesis) upon which a precise, concise and testable statement of the complete theory of evolution (including splitting of species) can be based."[7]

A crucial element in Williams' axiomatization is the definitions which she provides for her fundamental units, which she terms clans, Darwinian subclans, and subclans. "A clan will be defined as a set containing all of the descendants of some collection of 'founder' organisms; the class contains not only all contemporary descendants but also all descendants in all generations between the founder organisms and contemporary organisms." A Darwinian subclan is a portion of a clan "held together by cohesive forces so that it acts as a unit with respect to selection." Hence, Williams' theorems concern types of clans or Darwinian subclans as spatiotemporal wholes rather than as time slices of such wholes. She also limits her laws only to those groups which become established. Any "subclan" that does not become

6 "Deducing the Consequences of Evolution: A Mathematical Model," *Journal of Theoretical Biology*, 28 (1971), 343–85.

7 Ibid., pp. 343–44.

established cannot be counted as a subclan. For both these reasons, Williams is able to avoid the contingencies that follow upon Mayr's founder principle.

But, one might complain, evolutionary theory must deal with all those phenomena antecedently thought of as being evolutionary. Just as Newton's laws deal with all material bodies, Darwin's theory must deal with all living bodies. If something is alive, it must be included in some evolutionary unit. But scientific theories are quite often limited in their scope. For example, the genetical theory of evolution deals only with Mendelizing populations; i.e., populations made up of organisms exchanging genes in reproduction. At the very least, Williams' axiomatization shows which and how many of the phenomena antecedently thought of as the subject matter of evolutionary theory can be incorporated into a deductively organized theory of evolution.

Williams' axiomatization also points up the importance of the various levels of organization in evolution and how they are related. Genes, chromosomes, gametes, organisms, populations, and species all play different but related roles in evolution. A failure to see this is one of the major sources of the assertion that the principle of the survival of the fittest is a tautology. The only things that are passed on in reproduction are gene-bearing chromosomes. Heritable mutation is a necessary condition for evolution. It is the organism that interacts with its environment and either survives to reproduce or does not. Survival is a necessary condition for reproduction and reproduction is a necessary condition for the transmission of genes. But it is the population that is changed by this process. As Williams puts it, natural selection "is a strange kind of force, since the forces we usually deal with change the characteristics of the objects they act on; for example, an impressed force changes the state of motion of the billiard ball it acts on. Selection is a force which changes hereditary characteristics, but this change does not take place in individuals. Selection cannot change the inherited characteristics of an individual; it may, so to speak, punish the individual for his bad characteristics by preventing him from having offspring; but it is powerless to change his characteristics once he is there."[8]

A second important feature of Williams' treatment is that she explicitly takes note of the types of predictions that can be made on the basis of evolutionary theory—predictions not directly about particular individuals, properties of individuals, or even particular taxa, but about patterns of evolution. Evolutionary theory does not deal with the evolution of *Cygnus olor* but with the evolution of species. Numerous evolutionary patterns have been discerned by evolutionists, such as the invasion of an unoccupied ecological niche by an unspecialized species. To the extent that *Cygnus olor* happens to exhibit one of these patterns, evolutionists can make predictions about it on the basis of evolutionary theory. For long enough biologists have been abused

[8] Williams, "The Logical Status of Natural Selection and Other Evolutionary Controversies," *Philosophy of Science Symposium* (Dordrecht: D. Reidel, 1973).

for not being able to predict that all swans are white—especially because they are not. What they can and should be able to predict are truths about things like mimicry and protective coloration—regardless of the species that exhibit such evolutionary traits.

A traditional example of the type of predictions that can be made on the basis of evolutionary theory concerns superiority of the heterozygote, for instance at the hemoglobin S locus in man in areas in which falciparum malaria is endemic. Individuals homozygous for the normal hemoglobin allele are more prone to get malaria than are those who are heterozygous, and individuals homozygous for the hemoglobin S allele are more likely to die of sickle cell anemia. The principles of Mendelian genetics, enshrined in the Hardy-Weinberg law, can be used to determine various gene and genotype frequencies for a particular population, given the necessary information. The evolutionary prediction, however, concerns what happens when the selection pressure is removed in situations such as these; for example, when groups of individuals are transported from areas where falciparum malaria is endemic to areas where it is not. In successive generations, the frequency of the hemoglobin S allele should drop sharply. That this change is occurring in man is irrelevant to the prediction as a prediction based on evolutionary theory. That the two diseases are malaria and sickle cell anemia is also irrelevant. The relevant factors are the superiority of the heterozygote in this area and the relative frequencies of the two alleles—and relative frequencies are properties of populations, not individuals.

NATURAL SELECTION OR THE SURVIVAL OF THE FITTEST

The most serious and persistent criticism of evolutionary theory throughout its history is that the principle of natural selection or the survival of the fittest is a tautology. In general, the task of deciding whether a particular statement in a scientific theory is or is not a tautology is not easy. For example, considerable controversy surrounds Newton's second law of motion on this score. Many authors feel that evolutionary theory can be dismissed because the principle of the survival of the fittest is supposedly a tautology. For some reason, similar claims are not made about Newtonian theory. In the case of evolutionary theory, however, the principle of the survival of the fittest can be interpreted as a tautology only at the price of robbing evolutionary theory of the chief means by which it can be tested.

Numerous biologists claim that natural selection *is* the differential perpetuation of genes. What makes a gene fitter? Increase in relative frequency in later gene pools. Because genes are not passed on separately but in integrated genotypes, more careful biologists reword this claim to read that natural selection *is* the differential perpetuation of genotypes. What makes a genotype fitter? Differential perpetuation in later generations. If the biologist is organismically inclined, he might argue that natural selection *is* differential reproduction. What makes an organism fitter? Leaving more offspring. All of these versions of the principle of natural selection have several features in common: each specifies a necessary condition in evolution by natural selection, each errs in treating that necessary condition as if it were sufficient, and each omits any mention of the causal mechanisms responsible for the differential perpetuation specified. The end result is that each of the three versions taken separately degenerates into a tautology.

Differential perpetuation of genes and genotypes is certainly necessary. Without genotypes there could be no phenotypes. Of course, without phenotypes there could be no genotypes either. Certainly all these differential perpetuations are the results of natural selection, but natural selection is a relation between individual organisms and their environments. Omitting reference to these interrelations deprives evolutionary theory of much of its empirical content. For instance, the earlier example about the relative frequencies of the genes responsible for resistance to falciparum malaria and inclination toward sickle cell anemia would be meaningless without reference to the environment.

At this point, the reader might be willing to concede the importance of the causal role of the organism-environment relation in evolution, but object that nothing has been gained because this version of the principle of the survival of the fittest is equally tautological. One organism might survive to reproduce because of one character or set of characters in its environment and another because of a different character or set of characters in another environment, but the net result would still be that those organisms that survive to reproduce are *ipso facto* fit and those organisms that are fit *ipso facto* survive to reproduce. Degree of fitness enters in only with respect to the number of fertile offspring produced.

Several empirical objections can be raised to the above contention. For example, according to such a view all worker bees have the same coefficient of fitness—namely, zero. Some worker bees may be very efficient and hardy workers; others may consume more honey than they carry to the hive. But no matter. They are all equally unfit, because they are all sterile and leave no progeny. Only queens and an occasional drone pass on their genes in reproduction. Similar stories could be told for all sorts of social relationships, from one individual caring for the young of another to Hitler exterminating millions of Jews in World War II. There are two ways out of this empirical difficulty. One is to recognize these nonreproducing but efficacious individuals as part of the "environment" of those individuals that do pass on their genes. The other is to treat these groups of organisms as "individuals." For instance, the hive is the individual as far as evolution is concerned and the queen, drones, and workers the parts.

But even if all these and other empirical difficulties with the identification of "fitness" with "survival" are ignored, we are still left with the metaphysical compulsion that led us to the identification in the first place, and it *is* a metaphysical compulsion. In actual practice, fitness claims are made with respect to a particular trait or set of traits, not "all" the traits of the organism. One organism is fitter than another in a particular environment because it can withstand desiccation better than the other organism, and the dry season is upon them. Because the first organism has this trait more highly developed than the second, it has a better chance to survive. If one only knew enough about the genetic makeup, the embryological development, and the physiology of the organisms concerned, as well as the vagaries of the environment, one could assign a certain degree of fitness to each of these organisms and hence be able to make reasonable predictions about their chances of survival. With this information, one could in turn predict subsequent changes in the population.

The problem evolutionists have had to solve is the discovery of a level of evolutionary development which can be described in sufficiently broad terms to permit repetition of these events and the formulation of evolutionary laws, yet with sufficient precision to permit reasonable inferences from these laws. If the organisms and their environments are described too narrowly, the events become so particularized that they are likely to occur only once in the history of the universe, if at all. If the description is too broad, the laws are too loose to be of any service. As the situation now stands, laws concerning the evolutionary development of evolutionary units such as populations are not expressed in the form of true universal generalizations. Even to permit probabilistic predictions, they must be described with a high degree of precision. However, long before they have been described so extensively that they become unique, a level is reached at which additional descriptive precision does not result in any increase in the precision of the relevant predictions. The evolutionary development of particular species or populations as such cannot be predicted with any reasonable degree of certainty. Predictions are possible only to the extent that a population or species happens to fit one of the patterns of evolution which have currently been discovered.

One is tempted to rush by all such pragmatic considerations. Perhaps biologists do not know all the relevant variables and could not combine them meaningfully if they did, but surely nature does the summing for us. In principle, every organism that dies without leaving issue has a coefficient of fitness of zero. No matter that two individuals are identical twins with the same genotype—one could have an extremely high coefficient of fitness and the other a very low one, depending on how many offspring each leaves. The appeal to this retrospective deterministic bias is difficult to resist. If one individual dies without reproducing itself and another succeeds in leaving numerous offspring, something must have been responsible for the difference.

If only we had "complete" knowledge and an all-encompassing theory, the fate of every individual organism could be predicted with absolute certainty, and the survival of the fittest would become a tautology.

Little more can be said about the problem at this point, turning as it does on the cogency of such notions as a complete description, total knowledge, and the principle of universal causation. As far as I can see, the notion of a complete description presupposes a single, all-encompassing scientific theory for which the description would count as complete. The same can be said for total knowledge. Any stronger notions of complete description or total knowledge rest on highly questionable absolutist metaphysics. For example, it would seem that if one had total knowledge, there would be no point to science anyway. If one already knew everything, there would be no need to infer anything. In any case, evolutionary theory is not such an all-encompassing theory. As it now stands, the principle of the survival of the fittest is officially a tautology in certain operationally oriented versions of evolutionary theory, and these versions suffer accordingly. It is not a tautology in those versions of evolutionary theory which recognize the key role played in evolution by the organism-environment relation.

CHAPTER THREE

Biological Theories and Biological Laws

SCIENTIFIC LAWS AND DESCRIPTIVE GENERALIZATIONS

After all that has been said in the preceding chapters about genetic and evolutionary theories, it may seem somewhat gratuitous at this point to address ourselves to the question of whether or not there are any biological theories or laws. Yet several features of the theories and laws that we have discussed thus far conspire to lead some philosophers, for example, J. J. C. Smart (1963: 50), to argue that "there are no biological theories" analogous to the closely knit theories of physics; "there are not even any biological laws." But what of Mendelian genetics, molecular genetics, the genetical theory of evolution, the synthetic theory of evolution, Williams' axiomatization of Darwinian theory, and the theories of population biology of Levins, Lewontin, MacArthur, Wilson, and others? Are they not scientific theories? Do they not contain biological laws? Philosophers like Smart have found reasons to dismiss each of these proposed biological theories. To the extent that these theories possess the requisite coherence to be classed as genuine scientific theories, they are not truly biological. To the extent that they are truly biological, they are not genuine scientific theories.

Two issues seem implicit in most of the objections raised to the commonest examples of biological theories and laws. First, many of the examples of laws to be found in the biological literature are not process laws; that is, laws that permit the inference of all past and future states of the system, given the values for the relevant variables at any one time. Biological systems tend not to be sufficiently closed to permit the formulation of such laws. In biological contexts, other types of generalizations tend to take the place of process laws—causal laws, developmental laws, and historical laws. Certain authors tend to dismiss such generalizations out of hand. In this chapter we will investigate each of these putative kinds of biological law to see if it can count as a genuine scientific law. Even if one admits that process laws are in some sense the ideal toward which we should be striving, there still seems to be ample justification for investigating laws that fall short of this ideal, especially when it seems that they will be part and parcel of science for the foreseeable future. Even if biological theories and laws turn out to be less than perfect, some are less "imperfect" than others. There are good reasons for analyzing and comparing these "imperfect" creations.

Even if one grants that laws other than process laws can count as genuine scientific laws, a second major difficulty remains—distinguishing between natural laws and descriptive generalizations. It may be true that none of the first seven presidents of the United States of America were Christians (they were deists), but this is hardly a law of nature. As important as this distinction is in philosophy of science, the appropriate cut is not easy to make. Numerous criteria have been suggested to mark this felt difference. However, the only one that shows any promise of being adequate is the actual or eventual integration of natural laws into theories, while descriptive generalizations remain isolated statements. For example, both Bode's law concerning the distances of the various planets from the sun and Kepler's laws concerning their paths and velocities were at one time isolated generalizations limited specifically to a single star system. Bode's law, however, was not generalizable and is no longer considered a genuine law. The relative distances of the various planets from the sun is considered an accidental or historical feature of the solar system. That is, this feature of the solar system depends on the peculiar circumstances leading to its formation. There is no reason to expect these circumstances to be typical in the formation of star systems. Kepler's laws had the opposite fate and were integrated into Newtonian theory.

At any one time in the development of science it is not easy to decide which apparently true universal generalizations are merely descriptive or accidental and which will become incorporated into currently emerging theories. For example, all the major planets in our solar system happen to revolve in the same direction around the sun. All of them except Uranus and Venus rotate in the same direction relative to their planes of revolution. There is currently no reason to expect this to be true of all star systems.

Similarly, all the proteins that make up terrestrial organisms are of the same kind—the levo form. (A very few dextro amino acids are associated with certain bacteria.) Although there are chemical reasons for all the proteins in a single organism to be of the same form, there is nothing about current chemical or biological theory that would lead one to expect this form to be levo rather than dextro. From the point of view of current scientific theory, this fact is purely accidental.

Finally, we must explain our use of such words as "accidental" and "chance." Like "individual," these words will be used in a relative sense. At any one stage in the development of science, certain regularities lie beyond the scope of any scientific theory. Relative to these theories, these regularities are accidental features of the universe. Thus, an event will be random only with respect to a certain body of laws. With respect to a more inclusive or more powerful body of laws, it may be determined. An apocalyptic determinist, of course, is sure in advance that all events are caused and, hence, that there are no chance events. The existence of coincident chance events will prove to be important for historical inferences. For example, the fact that the proteins that make up terrestrial organisms are all of the levo form provides excellent evidence for the conclusion that all terrestrial creatures evolved from a single source. If the incorporation of levo rather than dextro proteins followed from current scientific theory, no such inference could be made. Coincident events help us to reconstruct the past only when we take the coincidence to be chance.

CAUSAL LAWS

The general heading under which all the laws to be discussed in this chapter fall is that of causal laws. This designation is not to be used here, as it frequently is, to refer to deterministic laws, but to generalizations in terms of causes, effects, and similar cognate terms. In their most perfect form, causal laws differ from process laws only with respect to time. Causes by definition never follow their effects. Process laws embody no necessary temporal asymmetry. The following discussion will be in terms of causal laws, not because I believe that such terminology is especially basic to science, but because departures from perfect knowledge can be discussed more easily in this vocabulary. Process laws deal with closed systems; causal laws are couched in terms of strands selected from an open web of causal connections. The basic requirement for causal laws is that natural phenomena be grouped into kinds that are contingently connected. The occurrence of one instance of one of these kinds can be used to infer the occurrence of an instance of another kind. The nature of these connections, however, and the consequent inferences vary. For example, a cause (C) may be both necessary and sufficient for its effect (E). From the occurrence of C, one can deductively infer E and vice versa. In such cases, a causal law can be replaced by a process law, losing only the temporal implication of the cause-effect terminology. As we have observed earlier, examples of such universal connections in biology are not easy to come by. However, one might say facetiously that being born is both necessary and sufficient for eventual death.

Although philosophers usually define causal connections in terms of universal correlation, the terminology of causes and effects can be retained for cases in which the correlations are not reciprocally universal. Sometimes causes are sufficient for their effects but not necessary—here, from the occurrence of *C,* one can infer *E,* but not vice versa. For example, the explosion of a star would be sufficient for the destruction of all life on one of its planets. As ecologists tirelessly remind us, it is hardly necessary. Conversely, sometimes a cause is necessary but not sufficient for its effect—from the occurrence of *E,* one can infer *C,* but not vice versa. Most cases of communicable diseases are examples of such connections. For example, the presence of the tuberculosis bacillus is necessary but not sufficient for contracting tuberculosis. Usually, however, causes are neither necessary nor sufficient for their effects—in these cases, from *C* we cannot deductively infer *E,* and from *E* we cannot deductively infer *C.* Yet *C* and *E* are not unrelated. A failure to recognize such formulations as legitimate instances of causal claims has permitted certain scientists (usually hired by the tobacco industry) to claim that smoking is not a cause of lung cancer. After all, many people smoke and never contract lung cancer, and many people contract lung cancer and have never smoked. But as reputable scientists are quick to point out, smoking and lung cancer are hardly unrelated. If "cause" is defined in terms of sufficient conditions or, worse, necessary and sufficient conditions, then we can never hope to discover the cause of lung cancer.

Many objections have been raised against calling generalizations of the types enumerated above "laws." If universal correlations have not yet been discovered, then the natural phenomena must not have been analyzed properly. Additional factors must be appended to necessary conditions to make the whole set sufficient. Presence of the tuberculosis bacillus plus low resistance, and so forth, is sufficient for contracting tuberculosis. Or the effect must be analyzed into subkinds. If a single cause cannot be found for a disease, subdivide it into several different diseases and see if perhaps a single causative agent can be found for each. Sometimes one or the other of these strategies is successful, sometimes not.

Another possibility is to provide a more sophisticated analysis of causation. Michael Scriven[1] suggests, for example, that typically a cause is a nonredundant member of a set of conditions sufficient to bring about the effect. Alternative sets of conditions are sufficient for the effect, but given *this* set,

[1] "Explanations, Predictions, and Laws," *Minnesota Studies in the Philosophy of Science,* III (Minneapolis: University of Minnesota Press, 1962), 215.

this particular condition is necessary for the occurrence of the event. There may be more than one member of the set which is necessary. Which of these is chosen for special mention as *the* cause of the event is context-dependent. Nor are we committed to being able to supply the complete set of conditions sufficient for the event. Typically we cannot. For example, someone might ask for the cause of John F. Kennedy's death. Numerous possible answers suggest themselves, depending on the intent of the questioner. Kennedy's death fits numerous reference classes—the death by gun wounds of a human being, the assassination of a head of state, and so on.

Those who object to calling the statement of such context-dependent, partial correlations "laws" maintain that they are characteristic of ordinary discourse and low-level scientific investigations; once natural phenomena have been adequately investigated, universal correlations will be forthcoming. Two replies can be made to this contention. First, in one intensively investigated area, the relation between genotype and phenotype, the requisite correlations have not been forthcoming because they do not exist. The connection between a particular gene and a particular gross phenotypic trait is on a par with that between smoking and lung cancer. Seldom is a particular gene necessary for a particular phenotypic trait and never is it sufficient. Hence the nature-nurture controversy. If a gene is not sufficient and/or necessary for schizophrenia, then it does not "cause" schizophrenia. Second, even if more intensive investigation does uncover correlations with the requisite universality, prior to this discovery knowledge of partial correlations can still be of some use. For example, it helps to know that smoking is correlated with lung cancer. Even after the discovery of possible underlying universal connections, for certain purposes the less universal but more easily obtainable gross correlations are still useful. For example, many geneticists still concern themselves with transmission genetics.

DEVELOPMENTAL LAWS IN ONTOGENY

One type of causal law extremely prevalent in biological contexts is that of a developmental law. In a single causal claim, one class of events is said to be the cause of a second. In a developmental law, a series of sequential events is specified. *A* occurs, then *B*, then *C*, and then *D*. *A* causes *B*, and *B* causes *C*, and *C* causes *D*. When the occurrence of each of these classes of events is both necessary and sufficient for the occurrence of the other classes, then developmental laws can be replaced by process laws in which *A, B, C,* and *D* are successive states of the system. Because the two-member causal chains in which the causes and effects are universally correlated are hard to come by in biological contexts, one would rightly expect the members of a developmental law to be correlated universally even more rarely. Examples of developmental sequences in biology are found most frequently in embryology, physiology, and paleontology. Embryology is concerned with the initial development of the organisms from the fertilized ovum to the adult, and physiology with the continued maintenance of the individual. Paleontology deals with the phylogenetic development of species. In each of these disciplines, developmental sequences are specified. The question is whether any of these generalizations, statistical though they may be, can count as scientific laws.

Many of the sequences discussed in the biological literature give the superficial appearance of being purely descriptive. For example, in the production of red pigment xanthommatin in *Drosophila,* tryptophan is converted to kynurenine, kynurenine to 3-hydroxanthommatin, and this is converted to xanthommatin. No one would object to calling statements characterizing each of these steps in this reaction scientific laws. From the point of view of each of these specific reactions, it is irrelevant that they are taking place in the cells of a particular species of *Drosophila.* Whenever the proper pH, temperature, catalysts, and so forth are present, tryptophan will be converted to kynurenine. Furthermore, such chemical laws are in principle derivable from quantum mechanics. However, one should not gloss too quickly over the prohibitively complex nature of the derivation of even the simplest characteristics of the simplest compounds and reactions. Many of the properties of individual water molecules can be computed by quantum-mechanical calculations, but physicists are still a long way from being able to derive the gross properties of large ensembles of water molecules from quantum mechanics.

A statement of each of the steps in the biosynthetic pathway would count as a scientific law, but what of the developmental sequence as a whole? The problem is not the lack of recurring instances. Time and again the developmental sequence specified above occurs. As an empirical generalization, it is well-founded. Nor is the problem that this biosynthetic pathway is necessarily limited to a particular species. Instances of various biosynthetic pathways can be found distributed throughout the plant and animal kingdoms. They could just as easily occur in nonterrestrial systems. What is needed before we can count statements of such developmental sequences as laws is the recognition of some system of laws from which they can be derived. To do this, we must enlarge our area of interest to include the genotype of the organism and how it functions to control this particular pathway. Stretching the phrase to its breaking point, it is in principle possible to predict the course of a biosynthetic pathway, given knowledge of the state of the organism, including its genetic program, and some general features of its environment. The derivation in this instance would be even more complex. Similarly, it is irrelevant that this reaction is taking place in the cells of a particular species of *Drosophila.* In any cell with the appropriate genetic makeup and environment, the biochemical reaction would proceed as programmed. However, knowledge of the genetic program would not be included among the *laws* in the explanation, but among the *boundary conditions*.

A similar story can be told for classical developmental biology in terms of the formation of blastulas, gastrulas, presumptive neural plates, mesoderm, and so forth. Such generalizations have in the past appeared to be merely descriptive both because they were formulated largely on the basis of observation and because they were not derivable in most instances from any well-formulated body of laws. In addition, the developmental series of gross embryology tend to be more limited in their distribution among various taxa than analogous biochemical sequences. But if the genetic makeup of a zygote is known and the environment of the developing organism controlled, then the occurrence of particular developmental sequences can in principle be specified. Some headway is being made in working out actual mechanisms in developmental embryology, but progress is proving to be slow and painful.

PHYLOGENY AND CROSS-SECTION LAWS

When we turn to developmental laws in phylogeny, the situation is significantly different from that in ontogeny, because nothing exists in phylogeny analogous to the genotype controlling phenotypic development. The existence of "evolutionary laws" is one of the most confused issues in the philosophy of biology. We have already discussed the confusion of evolutionary theory with phylogeny. The statement that certain species of Agnatha gave rise to the Placodermi and that certain species in this latter group gave rise to Osteichthyes is not part of evolutionary theory. It is a description of a particular phylogenetic sequence. The occurrence of this particular phylogenetic sequence was inferred with the aid of various scientific theories, including evolutionary theory, but the description of this sequence is just that—a description—and not an evolutionary law.

The problem that concerns us in this section is the apparent conflict between the demand that the terms which function in scientific laws be unrestricted spatially and temporally and the evolutionary definition of "species." Biologists from Aristotle and Theophrastus to G. G. Simpson and Libbie Hyman have grouped individual organisms into taxa of greater and greater inclusiveness on the basis of the covariation of phenotypic traits, usually morphological. For example, "Chordata" is defined in terms of a series of traits, including presence of internal gill slits, a dorsal hollow nerve cord, and a notochord during some stage in embryological development. According to the classical notion of definition, the set of traits listed in the definition must distinguish the species from all other species in the genus. All and only the members of this particular species have this particular set of traits. This is what is meant by saying that the characteristics used to define a term are severally necessary and jointly sufficient for membership. Other species might be characterized by each of the traits separately, but only in this species is this particular set found universally distributed among all members.

According to the traditional notion of definition, once a particular set of traits has been chosen as defining, statements of any additional correlations count as scientific laws, often termed cross-section laws. Process laws concern instantaneous changes or successions in time; cross-section laws concern contemporaneous correlations. At one time, "gold" was defined in terms of a set of properties including yellow color. Hence, "gold is yellow" would be an analytic claim. Assuming this definition, one might discover that gold is also malleable. Even though malleability is just as characteristic of gold as its color, the statement that "gold is malleable" would be synthetic and a cross-section law. Today, of course, "gold" is defined in terms of its atomic number. In the context of atomic theory, "gold has the atomic number 79" is definitional. All additional statements about the set of physical properties consequent upon an element having that atomic number are viewed as cross-section laws. The issue is whether similar observations can be made about statements like "all swans are white."

Previously, arguments were presented showing that species as evolving lineages cannot be delimited in the traditional manner. The names of particular species cannot be defined by a set of traits which are severally necessary and jointly sufficient for membership without excluding too many organisms from the evolutionary process. Instead the names of particular species must be defined in terms of properties which are characteristic of the species but which are not universally distributed among its members; that is, a set of "defining" characteristics is given, but no individual need possess all of them, only a sufficiently high percentage of the most important ones. Some traits are more characteristic of a particular taxon than are others, but no one character is strictly necessary. On this view of definition, the sharp distinction between defining characteristics and additional, empirically correlated characters breaks down. However, the difficulties that flow from the preceding considerations can be ignored here, because the issue is not how the names of particular species are to be defined but appropriate definitions of the term "species" itself. The distinction is the same as that between defining "physical element" and "gold." According to atomic theory, elements are distinguished in terms of their atomic number, the number of protons they possess. Gold has 79. "Element" is not defined in terms of the possession of 79 protons; "gold" is.

According to G. G. Simpson,[2] an evolutionary species is "a lineage (an

[2] *Principles of Animal Taxonomy* (New York: Columbia University Press, 1961), p. 153.

ancestral-descendant sequence of populations) evolving separately from others and with its own unitary evolutionary role and tendencies." The evolutionary definition of "species" applies to all types of species both sexual and asexual. The implication of the definition is that thresholds exist in nature. Species do not evolve like merging currents in the ocean, but separately, like branches of a river in a delta. As it stands, the evolutionary definition of "species" is not very "operational." Criteria are needed to facilitate its application. The biological definition of "species" provides one such criterion for sexually reproducing organisms. According to Ernst Mayr,[3] biological species are "groups of actually or potentially interbreeding natural populations, which are reproductively isolated from other such groups." Interbreeding is one mechanism for promoting genetic homogeneity, which in turn promotes evolutionary cohesiveness.

As evolutionary biologists have long contended, species as evolving lineages are spatiotemporal wholes, related temporally by the ancestor-descendant relation and spatially (among sexual forms) by gene exchange in reproduction. Because the chief criterion for deciding whether or not a particular complex is an individual, according to most analyses, is spatiotemporal continuity, it follows that species are individuals from the point of view of evolutionary theory. The ancient *Archaeopteryx* became extinct in the late Jurassic Period. If a group of organisms were to evolve that was identical in every respect to the ancient *Archaeopteryx* save ancestry, it would have to be recognized as a new and separate species. But as J. J. C. Smart (1963, 1968) has pointed out, if species are spatiotemporal individuals, their names are proper names and cannot function in scientific laws without violating the requirement that scientific laws be spatially and temporally unrestricted. Hence, statements like "all swans are white," even if true, cannot count as scientific laws.

Although Smart introduces unnecessary complications in his discussion by confusing the definitions of the names of particular species and definitions of "species," his argument can be reworked so that it is cogent. Both Smart and Michael Ruse[4] observe that if the evolutionary definition of "species" is abandoned and the names of particular species are defined in terms of

[3] *Systematics and the Origin of Species* (New York: Columbia University Press, 1942), p. 121.

[4] "Are There Laws in Biology?" *Australasian Journal of Philosophy,* 48 (1970), 234–46. In this paper Ruse counters many of the arguments raised by Smart against the existence of biological theories and laws, but he joins Smart in confusing definitions of "species" with definitions of the names of particular species like "*Cygnus olor.*" It is interesting to note that if species are to be treated as individuals, the difficulties presented for traditional notions of definition by the names of species as evolving lineages no longer obtain, because such terms become proper names which cannot be defined at all. But the implication of this decision for the taxonomic hierarchy are lethal, because species can no longer be viewed as being included in higher taxa. Some other analysis must be provided.

phenotypic traits independent of the evolutionary descent of the individuals involved, then the names of such species possess the generality necessary for functioning in scientific laws. The question that remains is whether these generalizations are scientific laws or merely empirical generalizations without theoretic backing. Ruse thinks that they count as scientific laws; Smart thinks otherwise.

Smart (1963: 235) admits that we can find sets of properties such that "so far as terrestrial animals are concerned, all and only mice possessed them. The trouble is that now we have no reason to suppose our laws to be true." As long as the evolutionary definition of "species" and its biological correlate are accepted, then "species" is a theoretical term in the amalgam of Mendelian genetics and classical evolutionary theory known as the synthetic theory of evolution. According to this theory, time slices of evolutionary species should form fairly tight clusters of individual organisms as far as their phenotypic characters are concerned. Once a particular species has been characterized statistically in terms of one set of traits, there is good reason to expect other unexamined traits to covary to some extent with those already examined. Abandon the evolutionary definition of "species" and its biological correlate and there is no reason to expect such covariations to hold here on earth, let alone throughout the universe.

"But," one might object, "if a warm-blooded animal lives in an environment in which the temperature falls too low, must not it have hair?" No—feathers, layers of blubber and heaven knows what else will do. At one time, ideal morphologists such as Cuvier believed that certain characters were necessarily correlated or discordant. For example, if an organism had feathers, then it could not have teeth. With the discovery of the *Archaeopteryx,* the duckbilled platypus and a host of other impossible combinations of phenotypic traits, biologists were forced to admit that at the level of phenotypic traits, there is always more than one way to skin a cat. Maybe all triangles *must* have three angles, but not all reptiles *must* have a three-chambered heart, though in point of fact they might. When fresh water vertebrates invaded salt water, they had to find ways to maintain the water content of their bodies while eliminating nitrogenous wastes. The methods they adopted depended in large measure on the peculiarities of the evolutionary situations in which they found themselves. If the evolutionary definition of species is abandoned, then statements referring to particular species must receive their theoretic backing from somewhere besides evolutionary theory.

In the early pages of this chapter, genuine scientific laws were distinguished from empirical generalizations by the actual or eventual incorporation of laws into theories. It is certainly true that gold is malleable, but this empirical generalization might have proved to be an accidental feature of the universe, like all the planets in our solar system revolving in the same direction around the sun. It did not. Now we can derive the properties of a physical element from its atomic structure and quantum theory. Statements like "gold is malleable" are extremely low-level consequences of quantum theory but they are still consequences. Can we do the same for similar cross-section laws in biology? If evolutionary theory does not permit us to infer such correlations, what will?

The only answer remaining is developmental embryology expanded to include knowledge of the genetic makeup of the organism. Any organism with the genetic makeup G in any environment ranging from E_1 through E_n, undergoing biochemical reactions R_1 through R_n, will come to have phenotypic characters $C_1, C_2, \ldots, C_n$. Thus, from knowledge of the genetic makeup of a class of organisms, the range of environments to which the members of this class are exposed and the biochemical reactions taking place, it is possible to infer that a certain set of phenotypic characters will result. However, careful notice must be taken of what must be known if such claims are to be warranted and of the extremely hypothetical nature of this inference. Currently we are even less able to deliver on such derivations than we are in the quantum theory examples.

Essentialists have argued that the names of species as evolving lineages cannot function in biological laws because the names of such species can be defined only in terms of statistically covarying characters. Smart and Ruse have argued that the names of species as evolving lineages cannot function in biological laws because such species are individuals, and scientific laws can refer only to classes. Smart has further argued that statements concerning the covariation of phenotypic characters in the absence of the evolutionary definition of "species" are not biological laws because there is no reason to suppose them to be true. If true, they would be merely empirical generalizations concerning accidental correlations. In this chapter I have argued for an extremely weak thesis, somewhat weaker than that argued by Ruse. Statements of the correlation of phenotypic characters can be counted as low-level biological laws but only when expanded to include reference to the genetic makeup of the organisms and the biochemical reactions that produce them. In point of fact, very few cross-section laws in biology possess such backing. Most of them tend to be of the "all swans are white" variety.

DEVELOPMENTAL LAWS IN PHYLOGENY

Whether species are viewed as evolving lineages or as open classes, empirical generalizations are not possible for the phylogenetic development of particular species. If species are treated as being evolving lineages, then the repetitions required for such generalizations are necessarily ruled out. The same evolutionary species can evolve only once. But even if species are not treated as evolutionary lineages, it is extremely unlikely that the "same" species would evolve with sufficient frequency to permit the formulation of empirical generalizations about its evolutionary development. Developmental laws in paleontology, if they exist at all, must be formulated at a higher level of generality than that of particular species. They must be formulated in terms of classes—kinds of species, populations, demes, Darwinian subclans, fitness sets, and the like. In point of fact, developmental laws in paleontology do not make reference to particular species like opossums, dogs, and crows, but to kinds of taxa like primitive, polytypic, and cosmopolitan species. Thus, evolutionary laws will not be of the form "all swans are *X*" but of the form "all species of type *Y* are *X*."

Numerous generalizations concerning patterns of evolutionary development can be extracted from the paleontological literature. Rensch lists an even hundred.[5] E. D. Cope is given credit for two such laws of evolution: his first law states that unspecialized species tend to avoid extinction longer than specialized species and his second law states that body size tends to increase during the evolution of a group. Dacqué's law states that contemporaneous species living in the same area tend to change in analogous ways. Williston's law is somewhat more complicated. It states that in forms that possess numerous parts that perform essentially the same function, the number of these structural parts will become reduced as these forms evolve and the remaining parts will become more specialized. Finally, Louis Dollo is given credit for the frequently made observation that evolution is irreversible; that is, evolving groups do not retrace the stages of their evolutionary development to return to an earlier state. An organism that has managed to leave the water and live on land may once again return to an aquatic life, but the two aquatic forms will differ markedly from each other.

Paleontologists use rules of thumb like those listed above to help reconstruct phylogenetic sequences. Like clues in a murder mystery, no one alone is very compelling, because numerous exceptions exist for all but Dollo's law. But if several seem to be reflected accurately in the fossil record, the concordance lends some credence to the reconstruction. These "evolutionary laws" have been listed, not because they are extremely accurate or because I wish to argue for their being genuine evolutionary laws, but to show that they do not contain reference to any particular species. They refer only to types of species. Hence, Smart's objections concerning species as evolutionary lineages are irrelevant. But what right have we to expect them to be true? Ernst Haeckel and O. H. Schindewolf, among others, have argued that species go through life cycles like individual organisms. We know this putative law is clearly false, because such trends are not discoverable in the fossil record. Nor do we know of any mechanism that could produce such cycles. But how are we to distinguish between those statements listed above

[5] *Biophilosophy* (New York: Columbia University Press, 1971).

that are genuine scientific laws and those that are statements of accidental coincidences? They all started as empirical generalizations. We must concern ourselves now with their theoretic justification.

Of all the laws listed, Dollo's law comes closest to being true and has the greatest theoretic support. Given the nature of genetic change and the retention of these changes in the genome, the likelihood that these changes would become totally erased and the genome returned exactly to an ancestral state is next to nil. A land-dwelling creature might take to the sea again, but it would retain vestiges of its former self which would not be present in its original aquatic ancestor. Dacqué's law sounds vaguely reasonable. Whenever all the organisms in an area are exposed to broad changes in their environment, one would expect them to adapt to these changes, but not necessarily in similar ways. All plants living in an area that is becoming more arid will have to find ways of conserving moisture, but the ways are very likely to be quite different.

In the main, however, it must be admitted that the theoretic justification for these putative laws is not great. They seem to be roughly on a par with the previously discussed cross-section laws and for the same reasons. The justification of cross-section laws concerning the correlation of gross phenotypic traits depends on knowledge of the genetic makeup of the organism and its response to variable environments—knowledge we rarely possess. Similarly, the generalizations listed above are formulated in order to permit long-range inferences about evolutionary development, given an absolute minimum of the requisite knowledge. No one should be surprised that they turn out to be none too accurate. If any genuine evolutionary laws are to be had, they are more likely to be of the type being suggested by modern population biologists. These putative laws concern the relation between such variables as convex and concave fitness sets (whether the environmental range is smaller or larger than the tolerance of the group), fine-grained and coarse-grained environments, and a variety of phenotypic strategies. (See Levins, 1968, and MacArthur, 1972.) Currently these evolutionary laws are of limited scope and tend to exist in isolated, partially inconsistent clusters. They have yet to be integrated into a single unified system.

HISTORICAL LAWS

One frequently reads that biology is unique because of the prevalence within it of historical laws. In fact, many of the various types of generalizations already discussed have been termed indiscriminately "historical laws." On the face of it, it is not easy to decide what it is that makes a law historical. A law cannot be termed "historical" just because it permits inferences about the past, because process laws do that. Nor does a law become "historical" merely because knowledge of the past is used in inferring the present or the future. Process laws can be used to do the same. Two types of generalizations, however, can be found in biological discourse that have some right to be termed historical laws. The first is a law that permits inferences *only* to the past. Evolutionary theory seems to permit explanation of past events in circumstances that preclude prediction. Such inferences will be discussed in the next section in conjunction with the thesis that scientific explanation and scientific prediction are symmetrical. Gustav Bergmann[6] has carefully analyzed the second type of generalization termed an "historical law," in the context of explanations in psychology, but much of what he has to say is applicable equally to biology.

According to Bergmann, an historical law is a generalization in which knowledge of the past is *necessary* to predict the future. Knowledge of the present alone will not do. For example, by merely looking at a cat sitting on the mantle, it is impossible to predict with any degree of certainty whether it will eat food placed before it. It might not be hungry. But if we know the cat is perfectly healthy and has been deprived of food for several days, we can predict with a high degree of certainty that it will eat the food placed before it. "Hungry" in this context is an historical concept, entailing something about the past history of the organism, and the statement that "hungry cats eat when presented with food" is an historical law. On Beckner's (1959) definition, an historical law is any law that contains an historical concept like "hungry." But "hungry" can also be interpreted as referring to the current physiological and/or psychological state of the organism. According to this interpretation, the concept and the corresponding law cease to be historical. (Those familiar with the controversy between behavioral, physiological, and mentalistic psychologists should be familiar with the general course of this argument.)

There is little doubt that many historical concepts and historical laws are replaceable by concepts and laws that are not historical in character. The important question is whether they are always replaceable. The contrast between historical and nonhistorical concepts and laws can be seen most clearly in genetics. If we limit ourselves just to gross transmission genetics, terms like "pure line," "heterozygote," and "hybrid" are historical concepts. All "pure line" means is that the stock has been inbred for numerous generations without the appearance of any novel states of the character being followed. "Hybrid" as an historical concept entails the crossing of two distinct stocks in the recent past. "Heterozygote" and "homozygote" can also be given historical definitions, but these definitions must be *disjunctive,* because several alternative past histories can give rise to a heterozygote or a homozygote. However, when we leave the phenotypic level and start discussing the fine structure of the gene, such terms as "homozygote" and "heterozygote" lose their historical reference. At bottom, an organism is

[6] *Philosophy of Science* (Madison: University of Wisconsin Press, 1957).

heterozygous for a particular trait because it is heterozygous for a particular gene, and the two alleles are heterozygous because they are structurally different. A geneticist could decide that two alleles were heterozygous without knowledge of their ancestry.

Throughout our discussion of genetics, we have ignored the historical dimension of the gene concept. It must be treated now. Homologous chromosomes are those chromosomes which pair at meiosis. In such cases, alleles are genes residing opposite each other on homologous chromosomes and, by an extension of the term, might also be termed homologous. But the word "homologous" is also used in a second sense to refer to evolutionary homologies. The notion of evolutionary homologies applies most directly to phenotypic traits. For example, the eustachian tube in mammals is evolutionarily homologous to the spiracle in sharks. Both developed from the same gill slit present in the earliest jawed vertebrates. In most extant mammals, the eustachian tube leads from the pharynx to the middle ear and serves to equalize air pressure on the ear drum. In most extant sharks, the spiracle is used to channel water from the outside to the gills. Although the eustachian tube in mammals and the spiracle in sharks differ from each other in both structure and function, they are evolutionarily homologous to each other because they can be traced back through series of ancestors to the same structure in the same ancestral group. Some biologists refer to any two structures derived phylogenetically from the same ancestral structure as being evolutionarily homologous to each other, regardless of how much structural change has taken place in the interim. Others recognize degrees of evolutionary homology, depending on how much change has taken place and how long ago the lineages split.

A similar story can be told for genes and chromosomes. By extending the meaning of the term, two genes can be said to be "evolutionarily homologous" to each other if they can be traced back through sequences of replications to a common ancestral gene. Similarly, two chromosomes are evolutionarily homologous to each other if they can be traced back through sequences of mitotic divisions to the same ancestral chromosome. However, in both of these cases, the frequency with which crossover occurs is liable to interfere with any simple identifications of evolutionary homologies at the levels of genes and chromosomes.

Biologists have been concerned primarily in deciding whether or not two apparently different structures are actually homologous to each other in an evolutionary sense. From the point of view of reconstructing phylogenies, it is crucial to distinguish similarities due to common ancestry from those that are due to convergence or parallel development. For our purposes the crucial question is whether or not these distinctions need to be made for the purposes of biological theories, especially evolutionary theory. Is it possible for two systems to be sufficiently similar to fall within the scope of a single law but behave differently because of their past history? Biologists are presented with elaborately organized systems. They cannot ignore this organization. This organization in turn is the result of a long series of evolutionary processes, some of which have left traces in the makeup of the system. These traces can be used to infer this past history, but do evolutionary homologies have to be distinguished from nonevolutionary homologies for the purposes of applying evolutionary theory?

The opinions of prominent biologists can be enlisted on both sides of the controversy. The disagreement rests on the role of history in biological theory. Two studies have been brought forward as being relevant to this question. In the first, Dobzhansky divided a single population of *Drosophila* into subpopulations and exposed each to similar environmental changes. They all responded similarly. Next he selected samples of various populations of the same species of *Drosophila* from different geographic locations, combined them to form mixed populations, and then exposed them to similar environmental changes. They responded differently. Dobzhansky concluded that the "outcome of such experiments depends upon the geographic origin of the ancestors of the experimental animals" (in Munson, 1971: 194). More accurately, the outcome depends on the genetic makeup of the ancestors of the experimental animals. Because populations from different geographic locations within the range of a single species are likely to have significantly different genotypic makeups, the response of each population to the same environmental change is likely to be different even without mixing them. Dobzhansky's experiments show not that biological systems are necessarily historical, but that predictions concerning the evolution of a particular species require knowledge of the genetic variation among the constituent populations of the species.

Richard Lewontin[7] has presented an even more sophisticated argument for the principle of historicity in evolution. When genetically identical populations are subjected to the same selection pressures but in the reverse order, "the two populations not only come to slightly different end points and have very different histories, but, more importantly, their average behavior has been totally different." Thus, Lewontin concludes, "The average genetic structure of a population in time depends not only on the static probability distribution of environments, *but on their historical sequence as well*. . .if the historical order in which environments occur is a significant variable in population adaptation, then an element of uniqueness is introduced." As ingenious as Lewontin's experiments are, they are perfectly compatible with these phenomena being law-governed and the replaceability of historical laws with nonhistorical laws, even process laws. If populations with identical genetic makeups are exposed initially to different selection pressures, one would expect them to respond differently. The fact that they are exposed to

7 "Is Nature Probable or Capricious?" *BioScience,* 16 (1966), 25–26.

the same selection pressures but in the reverse order is irrelevant. What has to be shown is that two populations that have different past histories but that have evolved sufficiently similar genetic makeups will respond to the same environmental changes differently because of their differing histories.

The point to notice is that if biological phenomena were "historical" in the sense just discussed, they would have been shown to be inherently acausal. "Complete" knowledge of the present would be inadequate to predict the future. Knowledge of the past history of the system would also be necessary. When the requirement for "complete" knowledge is dropped and we limit ourselves to just current theory, the implications of historical laws is not so drastic. Given current theory, it is perfectly possible for two systems to be sufficiently similar to fall within the scope of the same law and yet behave differently, not because of their differing past histories, but because of differences in their current makeup. Dobzhansky's experiment, described previously, shows how similar populations have to be to count as sufficiently similar if one wants to avoid alternative possible outcomes.

Max Delbrück (in Blackburn, 1966: 119) has argued that "any cell, embodying as it does the record of a billion years of evolution, represents more an historical than a physical event. . . You cannot expect to explain so wise an old bird in a few simple words." Simpson (1964) has observed that evolutionary theory differs from gravitational theory because "gravity has no history"; evolving species do. Ernst Mayr has repeatedly emphasized that living organisms are products of history and must be treated differently from inanimate objects. The message in the preceding quotations is certainly in part well taken. Living creatures have evolved. A legitimate undertaking in biology is the reconstruction of phylogenetic sequences. To do this, evolutionary homologies must be discerned, whether they occur at the level of genes, chromosomes or phenotypic structures. According to evolutionary theory, organisms can survive only if they exploit the energy resources of their environments with at least as much efficiency as their competitors. Hence, one would expect to find organisms to be highly efficient mechanisms, and for the most part they are. Occasional lapses in efficiency are explained in terms of the peculiar exigencies of origin. For example, in the human body, Henle's loop, the crossing of the respiratory and digestive systems at the pharynx, and the inversion of the retina hardly represent the best possible ways of performing the functions which they perform, but given what our ancestors had to work with at the time, it was the best that could be done—and it was good enough. (For further discussion, see Chapter Four.)

But once all of this has been said, we must still object to some of the possible implications of the preceding quotations. The reconstruction of phylogenies is a legitimate undertaking in biology, but so is the construction of scientific laws and theories. Gravity has no history, but neither does natural selection. Species have histories, but so do galaxies. Just as certain properties of particular species can be explained only in terms of the peculiarities of their history, certain properties of particular galaxies can be explained only in terms of their past histories. Cosmologists use much the same techniques in reconstructing past histories of galaxies that evolutionists use in reconstructing past histories of species. Galaxies retain fewer traces of their past histories than do species, but without such traces historical reconstructions would be impossible. So far no biologist has shown that genes, gross phenotypic characters, or species currently function differently *just* because of their past histories. Certain systems have the structures they do because of their past histories, but the only part of their past histories that matters is the part that has been incorporated into their current makeup.

To allay misunderstanding, a caveat must be inserted at this point. Like most philosophers, I have been concerned mainly with what is in principle possible and impossible. Scientists, on the other hand, are equally concerned with what can and cannot be done in actual practice, because they are the ones who must go about the task of doing science. I have argued that historical laws are in principle replaceable by laws concerning the present state of living organisms, including extensive knowledge of their genetic constitution. In actual practice, biologists do not have the knowledge necessary to do without their historical laws. They do not know the genetic makeup of the specimen before them. All they know is that it is the offspring of a white-eyed female and a red-eyed male. Much of contemporary transmission genetics is carried on without knowledge of or concern with the molecular structure of the genetic material or biosynthetic pathways. Even when an historical law is actually replaceable by a nonhistorical law, there is often no point in doing so. The historical law is good enough. When historical laws are useful, they should be used. When they are the only laws available, they must be used. In the preceding paragraphs I have argued that nothing except incomplete data stands in the way of replacing historical with nonhistorical laws.

THE COVERING-LAW MODEL OF SCIENTIFIC EXPLANATION AND PREDICTION

Why are nearly all swans white? Several answers can be given to this question depending on what is puzzling the person asking it. He might want to know just the phylogenetic history of swans. Usually, however, such an answer would not be enough, unless it was coupled with an explanation in terms of white being adaptive at some stage in the evolution of swans. Or the questioner might want to know why being white is currently adaptive. What is it that makes being black adaptive (or at least not maladaptive) for swans in Australia, whereas being white seems to be adaptive for swans elsewhere? Or he might want to know how the genome of white swans in response to their environments results in their being white.

On the surface, at least, the explanations produced by scientists differ significantly from each other. Several philosophers of science, whose views culminated in the classic paper by Hempel and Oppenheim (1948), have argued that a single analysis will do for all types of scientific explanation as well as for all types of scientific prediction—the covering-law model. The importance of this model for our purposes is two-fold. First, it provides us with a fairly self-contained example of philosophical analysis in science, and second, evolutionary theory has been invoked repeatedly in the arguments over the adequacy of the covering-law model, especially the thesis that explanation and prediction in science are in some sense "symmetrical," hereafter termed the symmetry thesis.

Philosophers of science as philosophers of science do not *do* science; rather they study science itself. Succinct as the preceding statement may be, it is ambiguous. In "studying" science, do philosophers merely describe what they find or do they, to some extent, legislate what science should be? Is the "science" they study actual scientific theories as historical entities or idealized reconstructions? Do they study just the products of scientific investigation or the process itself? Some historians claim to be passive recorders of plain facts; others admit that they do not merely describe. They select, interpret, organize, and reconstruct. Philosophers of science go one step further. They reconstruct and suggest changes or reinterpretations. At times, philosophy of science is indistinguishable from theoretical science. But downright legislation is also currently out of fashion in philosophy of science. Philosophers of science feel that they must argue for their reconstructions and suggestions and show that they are consistent with the "best" in science. What is considered "best" is, of course, partially determined by agreement with their reconstruction. Thus, the opponents of the covering-law model argue that it is inadequate because explanations and predictions in evolutionary theory do not fit the covering law paradigm, and the defenders argue either that they do or that evolutionary theory is inadequate because they do not. Such a manner of doing philosophy of science is often termed explication. In explication, the subject matter is actual science, but the philosopher does not feel obligated to leave his subject matter as he finds it. Perhaps scientists and other philosophers have not made a particular distinction, but they should. Much of this book has been explication.

The question of the proper subject matter of philosophy of science has already been touched on briefly. Some philosophers of science tend to concern themselves only with the logical features of scientific theories as they themselves have reconstructed them. None too surprisingly, they find in their reconstructions precisely what they themselves have built into the reconstruction initially. Others feel obligated to follow the historical development of scientific theories. At most they present only partial, temporary reconstructions. They also concern themselves with pragmatics. Methodological and epistemological questions are just as important as those that are more strictly logical. The controversy that has surrounded the covering-law model has been as much a function of differing philosophies of philosophy of science as of differing views on scientific explanation and prediction.

There are three elements of the covering-law model—one or more general laws $(L_1, L_2, \ldots L_r)$, a set of statements about particular circumstances $(C_1, C_2, \ldots C_k)$, and a statement of the event to be inferred from the former two sets of statements (E). The general laws in conjunction with the particular-circumstance statements form the *explanans*. The statement of the event being inferred is the *explanandum*. Together the whole argument schema, reproduced below, forms the *explanation* or the *prediction*. (See Hempel, 1965, and Rudner, 1966.)

$$\left.\begin{array}{l}\left.\begin{array}{c} L_1, L_2, \ldots L_r \\ C_1, C_2, \ldots C_k \end{array}\right\} \text{explanans} \\ \hline \left.\quad E \quad\right\} \text{explanandum}\end{array}\right\} \text{explanation}$$

In their original paper, Hempel and Oppenheim confined themselves to deductive explanations and predictions, containing only universal laws, though they recognized the existence of probabilistic explanations that contain at least one statistical law. Probabilistic explanations were to be treated in a later paper (Hempel, 1965). Certain advocates of the covering-law model argue that only deductive connections are sufficient for scientific explanation and prediction, and many of the early criticisms of the covering-law model were directed at this contention. This aspect of the dispute will be ignored in our discussion. Similarly, the particular circumstances mentioned above were originally termed "antecedent" conditions, but as the controversy progressed, the inappropriateness of this designation became clear. It does serve as a clue, however, to the original intent of the authors. Finally, the covering-law model applies to the inference of both singular and general statements. For the sake of simplicity, we will confine ourselves to the former. (Statistical generalizations can be deduced from statistical generalizations, but any inference from a statistical law to a singular statement is necessarily probabilistic.)

Before we turn to a discussion of the symmetry theses, a few words must be said about "ordinary usage." When Hempel and Oppenheim first set out their covering-law model with its accompanying symmetry thesis, their readers had every right to think that their analysis was more toward the descriptive end of the explication spectrum, but as the controversy progressed, the defenders have shifted more and more toward the legislative end of the spectrum. In the following exposition I do not propose to side with either strategy, though the general tenor of this book should indicate where my own preferences lie. The symmetry thesis, as introduced, appeared to be a significant, obvious relation between explanation and prediction in science. If nothing else, the critics of the symmetry thesis have shown that the significant relation is not obvious and the obvious relation is not significant.

THE SYMMETRY BETWEEN EXPLANATION AND PREDICTION

Limiting ourselves just to those explanations of single events that contain reference to general laws, "explanation" as it is most commonly used refers only to the *explanans*. In explanation, one knows that an event has occurred and proceeds to produce a set of laws and statements of particular circumstances from which the occurrence of the event can be inferred. One must have knowledge of the laws, the particular circumstances, and the event. In the most uncontroversial and strictest sense of the word "explain," one cannot explain an event that has yet to happen with laws that one does not possess by reference to unknown circumstances. "Explain" as it is commonly used has an epistemological commitment. Certain options are left open, however. The event being explained can occur before, in the midst of, or after the particular circumstances, but both must occur before the time at which the explanation is offered. This is the sense in which particular circumstances are "antecedent" in explanations. "Prediction" is even more restricted in its usage. A prediction is a statement to the effect that an event in the future will occur. In the paradigm sense of the word "predict," one cannot predict what has already happened. Predictions need not and usually do not include grounds for the prediction. In a more peripheral sense, predictions need not even have grounds or be the result of inferences (for example, prophecies). When predictions do have grounds and are the result of inferences, the relative time of the occurrence of the particular circumstances is more limited than in explanation. Because one cannot predict an event after it has happened and because one cannot infer from unknown circumstances, the particular circumstances must occur prior to both the making of the prediction and the occurrence of the event being inferred. (A scientist might be able to predict the particular circumstances on the basis of another set of particular circumstances, but eventually recourse must be made to current or past circumstances.)

In their original paper, Hempel and Oppenheim (reprinted in Hempel, 1965) claimed that the same formal analysis "applies to scientific prediction as well as explanation. The difference between the two is of a pragmatic character. If E is given, i.e., if we know that the phenomenon described by E has occurred and a suitable set of statements $C_1, C_2, \ldots C_k, L_1, L_2, \ldots L_r$ is provided afterwards, we speak of an explanation of the phenomenon in question. If the latter statements are given and E is derived prior to the occurrence of the phenomenon it describes, we speak of a prediction. It may be said, therefore, that an explanation is not fully adequate unless its *explanans,* if taken account of in time, could have served as a basis for predicting the phenomenon under consideration." And in a footnote, "The logical similarity of explanation and prediction, and the fact that one is directed towards past occurrences, the other towards future ones, is well expressed in the terms 'postdictability' and 'predictability' used by Reichenbach [*Quantum Mechanics*], p. 12."

Seldom have so few words uttered so casually by philosophers of science been exposed to so much scrutiny. If Hempel and Oppenheim could have foreseen the furor these remarks would engender, they certainly would have expressed themselves more carefully. Obviously, if explanation and prediction are to be in any sense symmetrical, we must depart from ordinary usage on numerous counts. A certain amount of explication is necessary. As it turns out, the only way the symmetry thesis can be maintained is to redefine these terms in ways seldom used by anyone. This does not detract from the possibility that this thesis, nevertheless, marks an important relation inherent in scientific inferences.

There are three elements in the covering-law model—the laws, the particular circumstances, and the event being inferred. Scientific laws, since the adoption of uniformitarianism in geology, have been held to be timeless. Hence, they do not "occur" at any particular time, but the other elements do. E can occur before, in the midst of, or after $C_1 \ldots C_k$. In addition, in order to *use* arguments exemplifying the covering-law model, the scientist must have knowledge of some of the elements. He can know the laws and that $C_1 \ldots C_k$ have occurred but not that E has occurred. He may know that E and $C_1 \ldots C_k$ have occurred but not know the laws. And so on for six more permutations. In their discussions of the relation between explanations and predictions, the authors involved were not always careful in specifying which of the possible states of affairs they were referring to. For example, postdiction and prediction are now distinguished from explanation and prediction, as well as retro-diction and prediction, not to mention retrodiction, pro-diction, and retroduction.

Grünbaum[8] attempts to short-circuit the criticisms of the symmetry thesis by contrasting assertibility with inferability. Whether or not a particular scientist is justified in asserting a particular explanation-argument or prediction-argument depends on all sorts of epistemological considerations. But the symmetry thesis concerns inferability, not assertibility. The criterion for distinguishing between an explanation-argument and a prediction-argument is the temporal relation of E to the time at which the argument is presented.

8 "Temporally Asymmetric Principles, Parity between Explanation and Prediction, and Mechanism versus Teleology," *Induction: Some Current Issues* (Middletown, Conn.: Wesleyan University Press, 1963).

An argument that fulfills the requirements of the covering-law model and is presented prior to the occurrence of the event being inferred is a prediction; afterwards, an explanation. Epistemological considerations such as the scientist's knowing that *E* has occurred are irrelevant to the symmetry thesis, according to Grünbaum's interpretation. A deductive argument presented before the occurrence of the event being deduced is just as deductive as the same argument presented afterwards. Identical observations can be made for inductive arguments. Rudner (1966: 63) concurs. "Whether a scientist's use of the argument will constitute an explanation or a prediction will depend on whether that use occurs at a time earlier than the event described by the *explanandum* (in which case he is using it to effect a prediction of that event) or subsequent to the time of the *explanandum* (in which case he is employing it to effect an explanation of the event)."

As a claim about the atemporal nature of argument forms, the symmetry thesis is unexceptional. It is hard to imagine anyone's bothering to make special reference to it, let alone anyone else's objecting to it. The objections that were raised to the covering-law model and the symmetry thesis have stemmed from a basic dissatisfaction with the identification of explanatory power exclusively with strength of inference. There is more to explanation than this, especially when the covering-law model is expanded to include probabilistic explanations. Hempel's critics have argued that deduction does not capture much of the connotation of "explanation" as it occurs in either scientific or ordinary discourse, nor does induction as it is currently explicated. Rather explanatory power might be judged in terms of information transmitted or statistical relevance. The best statistical explanation is the one that assigns the correct likelihood to the explanandum. The probability of an event's occurring might be low, but if this is the correct likelihood, then this is the correct explanation.[9]

OBJECTIONS TO THE SYMMETRY THESIS

Many of the counter-examples to the covering-law model are intended to show the independence of explanatory power and strength of inference. They are put forth in an attempt to show that the covering-law model is leaving something out—exactly that "something" which the distinction between laws and accidental generalizations was designed to mark. One can infer that an event will occur or has occurred from a true accidental generalization just as readily as from a genuine scientific law, yet the former would not seem explanatory. Whatever warrant genuine scientific laws have is precisely that warrant which we desire in scientific explanations. The criterion for distinguishing scientific laws from accidental generalizations that has been adopted

9 Wesley C. Salmon, *Statistical Explanation and Statistical Relevance* (Pittsburgh: University of Pittsburgh Press, 1971).

in this book is the actual or eventual inclusion of scientific laws into scientific theories. The problem with so many of the examples set out for or against the covering-law model is that they are presented in isolation, and frequently it is this isolation that makes the explanations seem so unsatisfactory. However, the opponents of the covering-law model would hardly be content with our criterion for distinguishing between scientific laws and accidental generalizations. It too leaves something out. The best way to describe this something is to term it a causal mechanism. "Why did Henry the VIII die? Because he was born." Although the correlation between people being born and their dying is universal, it seems to leave something out.

This section will be concerned with objections to the symmetry thesis of the covering-law model rather than to the model itself. Because so many of the objections raised to this thesis have depended upon epistemological asymmetries and because scientists tend to be more sensitive to epistemological than to logical considerations, biologists have tended to side with the opponents of the covering-law model. Of the various objections raised to the symmetry thesis, only two will be discussed here—the asymmetry that usually exists between causes and effects and the almost universal asymmetry that exists between records of the past and "records" of the future. These two objections to the symmetry thesis have been selected for discussion because evolutionary theory has been cited as lending support to them. It will be argued here that these are significant objections to the covering-law model but that evolutionary theory is not especially relevant to them.

Gallie, Scriven, and Toulmin[10] have argued that on occasion the stipulation of necessary conditions which are not sufficient for their effects is in some sense explanatory. As we have mentioned previously, sometimes causes are necessary but not sufficient for their effects. It should also be remembered that the converse is equally true. Sometimes they are sufficient but not necessary. In both cases, there is an asymmetry between the inferences that can be made between causes and effects. The defenders of the symmetry thesis have argued that the statement of sufficient conditions is always explanatory, whereas the statement of necessary conditions is explanatory only to the extent that these conditions approach sufficiency. The example usually cited in this connection concerns the relation between syphilis (primary syphilis) and paresis (tertiary syphilis). Contracting syphilis is necessary but not sufficient for developing paresis. In fact, rarely do people who contract syphilis proceed to develop paresis. Usually they are cured or else die from some other cause before paresis develops. Now someone might ask, "Why did Henry VIII develop paresis?" and receive the answer, "Because he had

[10] W. B. Gallie, "Explanation in History and the Genetic Sciences," reprinted in *Theories of History* (Glencoe, Ill.: The Free Press, 1959); M. Scriven, in Munson, (1971); S. Toulmin, *Foresight and Understanding* (Bloomington: Indiana University Press, 1961).

contracted syphilis." This answer seems mildly explanatory, but on just the knowledge that a person had contracted syphilis, we could not predict that he would develop paresis. In fact, given traditional reconstructions of inductive logic, we would be forced to predict that he would *not* develop paresis. Given no knowledge of a person, we would be forced to predict that he would never develop paresis. Given the added information that he has just contracted syphilis, we would still be forced to predict that he would never develop paresis.

Defenders of the symmetry thesis have countered that explanations in terms of necessary but not sufficient conditions are on a par with the statement "Why did Henry VIII die? Because he was born." All necessary conditions that are not sufficient are the same; they are all equally unexplanatory. This defense of the symmetry thesis is too strong if the covering-law model is to be expanded to include probabilistic explanations and predictions. At the very least, some necessary conditions are more nearly sufficient than others. If explanatory power is to be identified with the degree to which the inference is warranted, then those necessary conditions that are more nearly sufficient should be more explanatory. The opponents of the symmetry thesis counter that there is more to explanation than this. The stipulation of certain necessary conditions is highly explanatory, and this explanatory power is at least partially independent of the degree to which the inference from necessary conditions to their effects is warranted. This particular objection to the symmetry thesis may, however, actually stem from a dissatisfaction with current reconstructions of inductive logic in which an inference is not warranted until its probability surpasses the .5 level. Increasingly more complete sets of conditions seem to be increasingly more explanatory, especially if some of these conditions are necessary, prior to their having attained this level.

The attraction the preceding objection to the symmetry thesis of the covering-law model has had can be seen more clearly in the opposite situation, when the cause specified is sufficient but not necessary. For example, the earth reversing its direction of rotation (*à la* Velikovsky) would be sufficient to produce a tidal wave. It is hardly necessary. In fact, rarely if ever are tidal waves produced in this manner. This argument fulfills the requirements of the covering-law model beautifully, but we seem to have the same asymmetry of inference. From the cause, we can infer the effect, but from the effect, we cannot infer the cause. If one accepts the analysis of acceptable causal talk set out previously and if this is the symmetry which Hempel and Oppenheim intended, then the symmetry thesis is clearly false. In causal laws in which the cause is necessary for its effect but is rarely accompanied by it, inferences to the past but not the future are warranted. Conversely, in causal laws in which the cause is sufficient for its effect but rarely occurs in conjunction with it, inferences to the future but not the past are warranted. The dilemma seems to be that either explanation in terms of conditions that are necessary but nowhere near sufficient is spurious or else the symmetry thesis is false. Defenders of the symmetry thesis have felt compelled to opt for the first alternative; opponents for the second.

Evolutionary theory is especially relevant to this dispute only if it can be shown that laws specifying necessary but not sufficient conditions are more prevalent in it than in other scientific theories. For example, Mayr has argued that the introduction of some barrier to gene flow is necessary if speciation is to occur in sexual species. It is not sufficient. Thus, if speciation has occurred, we can infer that in the past some barrier to gene flow must have been introduced. The converse inference is not nearly so warranted. Similarly, an invasion of an unoccupied adaptive zone by an unspecialized species is nearly sufficient for rapid and extensive adaptive radiation. It is not necessary. If an unspecialized species has invaded an unoccupied adaptive zone, we can infer with a reasonably high degree of certainty that sometime in the future it will undergo adaptive radiation. The converse is much less warranted. Now it is certainly true that conditions that are nearly necessary are easier to come by in evolution than those that are anywhere near sufficient, but this seems to be true throughout the domain of science and not especially true of evolution. Thus, on the interpretation of the symmetry thesis presumed in this discussion, evolution and evolutionary theory do not seem to be especially relevant.

We have argued that there is an asymmetry between causes and effects. According to the interpretation of the symmetry thesis set out by Rudner and Grünbaum, the asymmetry between causes and effects does not count in the least against the symmetry thesis of the covering-law model. On their view, the symmetry thesis concerns the logical relation between explanans and explanandum. The strength of this inference does not change, regardless of whether the argument is presented before or after the event stipulated in the explanandum has occurred. If knowledge of Henry the VIII's contracting syphilis will not allow you to infer paresis before he contracts it, it will not allow you to infer it afterwards either. If knowledge of the earth's reversing its direction of rotation will allow you to infer a tidal wave before it occurs, it will allow you to infer it afterwards. Thus, one can see how impervious to counter-example the symmetry thesis as interpreted by Rudner and Grünbaum actually is. (It should be mentioned that Hempel does not concur in this interpretation; he of course may be mistaken in not doing so.)

A second asymmetry which supposedly exists between explanation and prediction is epistemological and has already been hinted at. In order to make inferences about the past or the future, we must have knowledge of the relevant particular circumstances. Because of the special nature of some of these particular circumstances, we are able to reconstruct the past with greater certainty and in greater detail than we are able to predict the future.

Certain natural phenomena contain traces of their past and thus can serve as records. Few if any natural phenomena contain "traces" of their future, though advance indicators like rapid decrease in barometric pressure before a storm do exist. In this instance, the correlations are not accidental but effects of a common cause. The changes that cause a storm also cause the mercury in a barometer to go down, but obviously the drop in mercury did not cause the ensuing storm. One could predict the storm by means of the change in barometer reading but not explain it in those same terms. In general, there are no "records" of the future. The one possible exception to this claim is precognition, but it need not detain us here.

Given evolutionary theory as it now stands and knowledge of current states of affairs *independent of there being records,* biologists can infer past and future states of the system with equal facility. That is, they can infer which possible past states of the system under study could have given rise to the current situation as well as which possible states are likely to eventuate in the future. Though the strength of the inference in both cases is fairly slight, the number of past possibilities is no more restricted than the number of future possibilities—if we ignore the fact that certain present-day data are records. We cannot infer future changes in the weather very well, especially long-term changes. But neither can we infer, using these same rules and principles, past changes in the weather with any high degree of certainty if we ignore geological records as records. Without this knowledge we can no more infer past evolutionary occurrences than those in the future. This is equally true of controlled experiments in the laboratory with populations of living organisms. Given knowledge of the current state of a population and the controlled changes in its environment in the past, we can infer with a reputable degree of certainty the initial state of the population. But we can do the same thing for the future. Given knowledge of the current state of the population and how we are going to vary its environment, we can infer its future state with the same degree of certainty.

But when we recognize that certain current states of affairs are records, the situation changes drastically. The number of past possibilities is substantially reduced. Given the order and nature of the geological strata, fossil remains, current DNA sequences, and numerous other contemporary records of past events, evolutionary theory can be used to aid in producing rather detailed reconstructions of past phylogenetic sequences. If the same data were available about the future, comparable reconstructions could be provided for it. In point of fact, they are not. It is this fact that is responsible for evolutionary biologists' being able to explain (i.e., reconstruct) the past so much better than they can predict the future. But this fact does not entail that evolutionary theory is in any significant sense "historical." Any reconstructions of phylogeny must be compatible with evolutionary theory. Evolutionary theory in turn can be used to reconstruct phylogenies, but exactly the same observations can be made concerning Newtonian theory and cosmogony. Reconstructing phylogeny is an historical undertaking. Evolutionary taxonomy is also historical, because it is committed to representing phylogenetic development by means of a nested set of taxonomic classes. Evolutionary theory is not, thereby, an historical theory.

Rudner (1966: 60) has observed that in "the case of explanation, we have, so to speak, our *E*," the event to be explained. From the point of view of explanation-arguments and prediction-arguments set out by Grünbaum and Rudner, this difference makes no difference, but from an epistemological point of view, it does. As Scriven has emphasized, if an event has occurred, then all of the conditions necessary for it must also have occurred. Reference to these necessary conditions is, in certain circumstances and to some extent, explanatory. In addition, the event being explained may contain traces of its own history, further enhancing our ability to infer the occurrence of the events that preceded it.

UNIQUENESS AND NARRATIVE EXPLANATIONS

The covering-law model of scientific explanation has been challenged on several fronts. T. A. Goudge (1961) has argued that it is inadequate to account for a very special type of explanation to be found in evolutionary studies, which he terms narrative explanations. In a narrative explanation, an event like the evolution of mammals or extinction of dinosaurs is explained by specifying the temporal sequence of events which led up to it. Of course, the evolutionist does not mention everything preceding the event to be explained; instead he mentions the most important antecedent conditions. Sometimes these conditions are necessary or nearly so, and sometimes they are circumstances that are unexpected. Rarely if ever are they sufficient. If Goudge were arguing only that necessary conditions are occasionally explanatory in some pre-analytic sense of "explain," then we would have no quarrel with him, but he goes even further. He argues that in narrative explanations, no recourse is made to laws, either explicitly or implicitly. Goudge shows justifiable exasperation with the tiresome insistence of defenders of the covering-law model that all explanations refer to laws at least implicitly, even when these laws cannot be produced, but he does not stop here. He maintains that events in evolution are unique. Hence, explanation in terms of subsumption under scientific law is impossible. "Whenever a narrative explanation of an event in evolution is called for, the event is not an instance of a kind, but is a singular occurrence, something which has happened just once and which cannot recur. It is, therefore, not material for any generalization or law."

If we are to make any sense out of Goudge's contention, we must distinguish, as we did in Chapter Two, between unique events and necessarily unique events. A unique event is one that happens to be one of a kind. For example, the Catholic Church may someday elect an Irish pope. If it does, that will be a unique event but not *necessarily* unique because it could recur. The election of the *first* Irish pope would, however, be necessarily unique, because by definition it could not recur. The message of these examples is that events are not unique in and of themselves but only under certain descriptions. A necessarily unique event can be given a proper name and a definite description, but that is all. Hence, if we are to make sense of Goudge's position, he must be interpreted as stating that there are no significant kinds of events in evolution.

On the above interpretation of Goudge's thesis, we are left with two questions. What is it about evolution or evolutionary theory that seems to necessitate Goudge's conclusion that evolutionary events are necessarily unique? And in what sense does an historical narrative explain a unique event? When Goudge claims that evolutionary events are unique, he may have the principle of monophyly in mind. If all evolutionary events involve taxa and if taxa are necessarily monophyletic, then one might be tempted to argue that evolutionary events are necessarily unique. By definition, mammals as monophyletic units can evolve only once. Hence, the evolution of mammals is a necessarily unique event. But under different descriptions, the evolution of various types of taxa is no longer unique. For example, the evolution of mammals might be viewed as an instance of the invasion of an unoccupied adaptive zone. Such an event can and has happened more than once. In short, if taxa must be minimally monophyletic, then they are individuals and no direct reference can be made to them in scientific laws, but this admission does not entail that all evolutionary events under any and all descriptions are unique.

Usually when people speak of something's being unique, they do not mean "unique" in the strict sense described above. Rather they mean that the event has been characterized in such detail that it is extremely unlikely that a second instance will ever occur during the entire history of the universe. For example, in the following quotation G. G. Simpson (1964: 186) contrasts explanations in physics and the explanation of historical processes:

> Physical or mechanistic laws depend on the existence of an immediate set of conditions, usually in rather simple combinations, which can be repeated at will and which are adequate in themselves to determine a response or result. In any truly historical process, the determining conditions are far from simple and are not immediate or repetitive. Historical cause embraces the *totality* of preceding events. Such a cause can never be repeated, and it changes from instant to instant. Repetition of some factors still would not be a repetition of historical causation. The mere fact that similar conditions have occurred twice and not once would make an essential difference, and the materials and reagents (such as the sorts of existing organisms in the evolutionary sequence) would be sure to be different in some respects.

A rather straightforward interpretation of Simpson's remarks makes them applicable to all events, not just those in the evolution of species. Any event can be characterized so minutely that it is unlikely to recur. If "cause" is defined as the totality of preceding events, then no event can be explained by subsuming it under a general law. If a new group of organisms tried to occupy terrestrial ecological niches as reptiles did long ago, and even if all other conditions were the same, the situation would be different, because these niches are now occupied. But similar observations can be made for any physical system. Even if the solar system is viewed as a completely isolated system, it is extremely unlikely that all the solar bodies have ever in the history of the solar system occupied the same positions relative to each other more than once.

So far, we have not succeeded in interpreting Simpson's remarks so that they distinguish between physics and biology. Perhaps, however, he and Goudge have an even subtler point in mind. Accurate predictions can be made about planets, balls rolling down inclined planes, and such without characterizing these events in any great detail. The descriptions are short and simple, permitting numerous recurrences of the "same" event, yet the laws are universal in form and very accurate. In fact, one of the major tasks of scientists is to discover descriptions of the phenomena which they are studying that permit the formulation of such laws. The two go together. Geneticists like Dobzhansky, Mayr, and Lewontin have shown us how precisely the genetic makeup of a population must be specified if we wish to predict its future development with any respectable degree of accuracy. Simpson and Goudge may be asserting that this degree of specification is so great that no laws governing these events will be forthcoming.

Or they may be referring to the pecular imbalance that exists in evolutionary explanations between the particular circumstances and anything that might count as evolutionary laws. According to the covering-law model of explanation, reference must be made to both laws and particular circumstances. In the usual examples given of covering-law explanations, the laws are the chief explanatory elements. The specification of the particular circumstances, though necessary, is of no great significance. But in evolutionary explanations, the emphasis is just the opposite. It is the particular circumstances in the form of an historical narrative that seem to bear the brunt of the explanatory load. There may be some laws lurking about somewhere, but rarely are they mentioned, and when they are, they do not add much weight to the explanation. They tend to be of the form "animals with longer legs can run faster," simultaneously a truism and false. Proponents of the covering-law model of explanation would reply to this observation, "So much the worse for evolutionary explanations." Critics would reply, "So much the worse for the covering-law model of scientific explanation." The conflict is between our feeling that historical narratives are in some sense explanatory and the requirements of the covering-law model of explanation. In any case, if Simpson or Goudge means that evolutionary events are necessarily unique, then it is hard to see how an historical narrative could in any sense explain them.[11]

Several senses of "evolutionary law" have been distinguished in this chapter. Sometimes descriptions of particular phylogenetic sequences are referred to as evolutionary laws. If taxa must be monophyletic, then all such sequences are necessarily unique, and on no analysis of scientific law current in the literature can descriptions of unique sequences of events count as scientific laws. However, if we abandon the principle of monophyly, then phylogenetic sequences need not be unique, but any descriptions sufficiently detailed to permit the identification of particular taxa are liable to preclude all but the rarest repetitions of phylogenetic sequences. Hence, if such statements are to count as scientific laws, they must be inferable from evolutionary theory. The difficulties confronting such inferences were set out in Chapter Two.

Frequently the rules of thumb used to reconstruct phylogenetic sequences, such as Dollo's Law, are termed evolutionary laws. Since these putative laws do not refer to particular taxa, the principle of monophyly presents no problem. However, the theoretic backing for the hundred or so of these laws is not great. At best they are empirical generalizations of extremely restricted scope and with numerous exceptions. Causal laws in which the cause is necessary but not sufficient for its effect pose problems for the analysis of the concepts of both scientific law and scientific explanation. In such cases, one can infer the existence of the cause from the occurrence of its effect, but not vice versa. Once phylogenetic reconstructions are distinguished from inferences from evolutionary theory, this particular dispute turns out not to be especially relevant to evolutionary theory or evolutionary theory to it. In its most significant sense, "evolutionary law" refers to those statements which go to make up one or more of the versions of evolutionary theory sketched in Chapter Two. These laws are "evolutionary" in the sense that they refer to evolutionary processes, but they are not thereby in any significant sense "historical."

[11] See M. Ruse, "Narrative Explanation and the Theory of Evolution," *Canadian Journal of Philosophy,* 1 (1971), 59–74, for a more detailed criticism of Goudge's position.

CHAPTER FOUR

Teleology

THE PROBLEM OF TELEOLOGY

The chasm that yawns between the subject matter of the preceding chapters and the metaphysical doctrines of the early Greeks and later Christians would seem to be all but unbridgeable. In retrospect, these teleological metaphysics seem to be little more than weak analogies. In the case of Aristotle's internal teleology, the analogy was between the organization exhibited by living organisms and the universe at large, a functional analogy. In the case of the Christian, Neo-Platonic external teleology, the analogy was between a good man acting wisely and an all-powerful God governing the universe, a purposive analogy. But, one might object, evolutionary theory did away with teleology, and that is that. Yet the biological phenomena that gave rise to the idea of teleology are still with us, and biologists still continue to talk teleologically. Are biological phenomena so different from those of physics and chemistry that they require a radically different explanatory vocabulary? Teleological modes of expression are often more convenient than ordinary causal talk when describing the behavior and organization of living systems, but are they indispensable?

Thus, in spite of the apparent disparity between the subject matter of this chapter and of the previous chapters, the problems are not all that different. In Chapter One, a central issue was the proper definition of the gene. What are genes? Should the gene be defined functionally, structurally, historically, or in terms of its relative position on its chromosome? In this chapter, one of our chief concerns will be the proper definition of teleological systems. What are teleological systems? Should they be defined functionally, structurally, or historically? No one has suggested a "positional" definition of teleological systems, whatever that might be. Are there several different types of teleological systems with little in common but a name? Is there a single unified concept that possesses distinct varieties? Or is the notion of teleological systems a hodgepodge?

The problem of teleology can be approached empirically, by examining systems that are traditionally viewed as being teleological to see if some property or set of properties can be found characteristic of them all. Four such properties, usually grouped in pairs, have been suggested. The first pair consists of one functional and one structural definition. The functional definition is in terms of the frequency with which certain preferred states or goal-states of the system are attained in the face of a wide range of changes both in the system and in the environment. Its structural partner is in terms of the causal mechanisms that bring about such preferred states, causal feedback loops, especially negative feedback. The second pair consists of one historical and one structural definition. The historical definition is in terms of the origin of teleological systems through selection processes. Its structural partner is expressed in terms of programs.

The problem of teleology can also be approached linguistically. Both physicists and biologists use ordinary causal language in describing and explaining natural phenomena. Just as a physicist might say that heating a gas causes it to expand, a biologist might say that heating a mammal causes it to sweat. But a biologist might also say that a mammal sweats when heated in order to keep its temperature constant, while no physicist would say that a gas expands when heated in order to keep its temperature constant—even though that is exactly what happens. Hence, we are presented with a problem which is similar to, though more difficult than, that discussed in Chapter One, the reduction of Mendelian to molecular genetics. It is the reduction of teleological modes of expression to ordinary causal locutions. The problem is similar to that discussed in Chapter One, since we must work out the linguistic and logical relations between two conceptual schemata. It is more difficult because in the case of genetics we had two reasonably well-articulated scientific theories before us. In the reduction of teleological to ordinary causal modes of discourse, all we have to go on are two informal modes of expression. Sometimes biologists use teleological language in highly technical contexts, but the teleological discourse itself is not technical. Scientists use teleological language; they rarely analyze it. Hence, our conclusions about the reduction of teleological to causal modes of expression must necessarily be even more tentative than those reached about the reduction of Mendelian to molecular genetics.

In the following discussion, "teleological" will be used in a generic, metaphysically neutral sense to refer to all statements couched in terms of goals, purposes, and functions. "Teleonomic" would serve as well in most contexts, because our analysis eventually converges on that set out by the proponents of teleonomy.[1] "Functional" is reserved for those situations in which the organization of living creatures is paradigmatic, and "purposive" for those situations in which conscious human behavior is the paradigm. At times people do behave in a manner calculated to attain a goal. Occasionally other organisms and even certain artifacts behave in a like manner. The following discussion is couched in terms of teleological systems. It could be carried on as easily in terms of teleological processes, behaviors, or entities.

THE ESSENCE OF TELEOLOGY

In the following pages we will examine the criteria which seem most characteristic of teleological systems —the prevalence of preferred states, closed feedback loops, and programs, and the origin of such systems by means of selection processes. Taken together, these criteria seem to distinguish a class of systems which accords fairly well with our intuitive, pre-analytic notion of teleological systems, but the boundaries determined by these criteria do not always coincide. We cannot set out a list of criteria which are severally necessary and jointly sufficient for a system's being teleological. To no one's surprise, I am sure, we have not discovered the essence of teleological systems. On purely inductive grounds of the crudest sort, one should expect the search for the essence of teleological systems to be no more productive than the search for the essences of society, art, space, horses, and mankind. Many authorities would maintain that each of these notions has an essence and that they know what it is. Unfortunately, no two authorities can be brought to agree on exactly what these essences in each case actually are. It is difficult enough to define terms that occur in highly developed, tightly organized scientific theories by a single set of necessary and sufficient conditions; it would be truly amazing if something with as tortuous a history as the notion of a teleological system could be defined in this way.

The next step, of course, is to claim that "teleological system" is a cluster concept. No one set of conditions is severally necessary and jointly sufficient for membership, but several alternative, closely related sets of conditions can be used to characterize it. As true as this claim may be, it does not get

[1] C. S. Pittendrigh, in *Behavior and Evolution,* A. Roe and G. G. Simpson, eds. (New Haven: Yale University Press, 1958); and Mayr in Blackburn (1966) and Munson (1971).

us very far. It is easy to claim that a particular term is a cluster concept. It is quite difficult to list the relevant characteristics and specify precisely how they are related in order to decrease the halo of vagueness that surrounds the concept. To do so, first we will discuss two physical models that have been proposed for teleological systems, the temperature-control schema for functional systems and the torpedo schema for purposive systems. The reason for utilizing such models is that at least many of the characteristics of teleological systems can be set out in terms of such physical systems, using the ordinary laws of physics. Hopefully, then, the principles derived from the study of such simple systems can be extended to the structure and functioning of living organisms and even conscious agents.

Next we will turn our attention to the two pairs of conditions that have been suggested for defining the notion of teleological systems. How well do they handle paradigm cases of teleological systems? How well do they handle various borderline cases? If anything is to count as a teleological system, living organisms must. Any analysis that excludes them is certainly mistaken. Conversely, any analysis that includes the solar system is equally mistaken but for the opposite reason. Borderline cases present a different problem. They have some of the characteristics of teleological systems but lack others. How well a particular analysis handles such cases as the behavior of a compass needle, chromosomes lining up at meiosis, computers, and Old Faithful errupting every hour or so determines to some extent how good it is. The chief problem is finding a justification for the decisions we make.

For example, anyone observing the behavior of the child's toy known as the walking-beetle would be tempted to describe its behavior teleologically. When wound up and placed on the top of a table, it wanders around the surface, always turning aside when it starts to fall off the edge. The mechanism that produces this behavior is a flywheel spinning at right angles to the other wheels. When the front wheels of the toy start to drop off the edge of the table, the flywheel comes into contact with the surface of the table and the toy is directed away from the edge. In treating the behavior of such an artifact teleologically, are we merely reacting to its bug-like shape and decoration? Once we know the mechanism by which it works, would we still be tempted to treat it teleologically? Where is the recurrent preferred state, the network of causal feedback loops, the program, and the selection process that gave rise to it?

Answers to these questions are certainly forthcoming. The problem is to find some grounds besides pre-analytic intuitions to evaluate them. In attempting to decrease the vagueness that surrounds the notion of teleological system, one soon reaches a point where there is no reason to make one decision rather than another, no reason except animal prejudice. And throughout the history of philosophy, one man's self-evident truth has turned out to be another man's animal prejudice. It is not easy to decide what is to count as an electron, gene, or species, but decisions can be made and good arguments brought forth to defend them because these terms are embedded in scientific theories. The notion of teleological systems is not. That is why we will emphasize those criteria that have some foundation in current scientific theory, specifically evolutionary theory and learning theory. Perhaps neither of these theories is developed to the extent that one might wish, but at least they provide something more substantial to go on than our pre-analytic or even post-analytic intuitions.

Finally, in this chapter we will deal with three related problems in the philosophy of biology. How do teleological statements like "a function of the heartbeat in mammals is to circulate the blood" differ from ordinary causal statements like "the heartbeat in mammals causes the blood to circulate"? How do teleological explanations differ from ordinary causal explanations? And how does each of these kinds of explanation differ from the covering-law model of scientific explanation? Our overriding concern, however, will be to find some justification for whatever decisions we make.

PHYSICAL MODELS FOR TELEOLOGICAL SYSTEMS

Scientists tend to distrust analogies almost as much as they use them. The notable failures of functional analogies in science from the teleological metaphysics of Aristotle to functionalism in the social sciences have earned an especially bad reputation for functional analogies. But reasoning by analogy is one of the most successful tactics in the process of discovery, and most of the criticisms of functional analogies have been directed at the crudities of their use in the social sciences, not at biology (Hempel, 1965, and Rudner, 1966).

In reasoning by analogy, the behavior of a poorly understood system is assimilated to the behavior of a well-understood paradigm system. Hopefully the principles that govern the behavior of the paradigm system can be extrapolated to the poorly known system. Often the relation between the paradigm and the assimilated system is reciprocal. Initially knowledge of the paradigm is extrapolated to the assimilated system to aid in understanding it. Additional knowledge acquired about the assimilated system then helps in understanding the paradigm. Sometimes the center of gravity of the analogy shifts until the system originally chosen as the system to be assimilated becomes better understood than the paradigm system. In fact, this is exactly what has happened in modern biology and psychology. Scientists now feel that they know much more about the workings of the universe at large than they do about the structure and functioning of living organisms, especially the human brain. Instead of reasoning from the behavior of living organisms or conscious agents to purely physical systems, scientists and philosophers have tried to understand living organisms, including conscious agents, by means of physical models. It is to an analysis of such physical models that we now turn.

The temperature-control schema has been suggested as a model for functional systems, especially those that are homeostatic. In this model, the temperature of a house is maintained by means of a thermostat connected to air-cooling and heating units. For example, if the thermostat is set at 72°F and the air temperature inside the house rises above this level, the air-cooling unit is switched on until the temperature falls again, then shuts off. A similar chain of events follows when the temperature falls too far below the 72°F mark. Thus, even though the temperature does not remain exactly at 72°F, it oscillates around it. The torpedo schema has been suggested as a model for purposive behavior. In this model, a torpedo is fitted with a motor-driven propeller and a device for homing in on a sound source. A torpedo is aimed in the general direction of a target ship. When it picks up the sound being emitted from the engines of the ship, the steering mechanism directs the torpedo at the source of this sound. No matter how the target ship might maneuver, the torpedo continues to home in on it. At least, this is how it is supposed to work.

In certain respects, these models are typical of the systems they are intended to represent; in others they are not. In the temperature-control schema, a single variable is maintained within limits at the expense of changes in the parts of the system, but in some homeostatic systems, it is a state of the whole system as such that is maintained. For example, the genotype-phenotype relation can be viewed as an example of homeostasis—developmental homeostasis. In developmental homeostasis, a given genotype gives rise to the same phenotype under a wide variety of environmental conditions. Similarly, a population must remain adapted to its environment even though this equilibrium is constantly being interrupted by changes in the environment or the introduction of new genes into the population. The ability to make such adjustments is termed genetic homeostasis.

In homeostatic systems, either a single variable or a state of the whole system is maintained within limits under various, though not unlimited, changes in both the environment and the system itself. But another type of functional system also tends to be described teleologically—regular, sequential changes in time. At any one stage in the life cycle of an organism, numerous homeostatic mechanisms maintain its organization, but the organism also undergoes embryological development. Instances of such stabilized flow have been termed canalization or homeorhesis. In homeorhesis, instead of a preferred state being maintained, a series of preferred states are attained, usually through a narrowly prescribed path. The halter once used on horses which permitted them to increase the arch in their necks but once increased, never to decrease it, is a simple physical analog to such homeorhetic systems.

Homeorhetic systems are similar in many respects to the second main type of teleological systems—purposive systems. As numerous philosophers and scientists have argued, purposive behavior can be explained without attributing causal efficacy to future events. In the case of the torpedo, an ordinary causal chain leads from the engine of the target ship to the sound-detecting device and from there to the steering mechanism of the torpedo. The goal of the torpedo might be a future event, the destruction of the ship, but this future event is not determining the present behavior of the torpedo. The torpedo homes in on the ship—its goal-object—not on the future event.

One apparent dissimilarity between the temperature-control schema and the torpedo schema is the absence of anything like a goal-object in the former. But the torpedo schema is atypical of purposive behavior in this respect, because purposive behavior is frequently exhibited in the absence of any goal-object. For example, the goal of a prizefighter may be to win an upcoming fight, but until he climbs into the ring he has no target. A dog may chase a rabbit, but it also may sniff around looking for something to chase just as a homing torpedo may wander aimlessly until it picks up an appropriate signal. Homeostatic and homeorhetic systems can be assimilated to a common paradigm if homeorhetic systems are viewed as homeostatic systems going through a single cycle or homeostatic systems are viewed as homeorhetic systems going through successive cycles. Nothing but the mind-body problem stands in the way of doing the same for purposive behavior.

PREFERRED STATES AND NEGATIVE FEEDBACK

Numerous philosophers and scientists have tried to analyze the notions of preferred states and negative feedback, two of the commonest criteria used to distinguish teleological from nonteleological systems. However, these analyses are extremely complicated and utilize highly technical notations and vocabularies. Any attempt to paraphrase them while reflecting all the niceties of their formulations would render them unintelligible to anyone who had not already studied the originals. To make matters worse, none of them is very successful. Hence, all I intend to do in the following pages is to sketch two representative attempts at analyzing the notion of a teleological system in terms of preferred states and causal feedback loops, suggesting where and why each falls short. Anyone interested in more detailed discussions will have to return to the original literature.[2]

One way of distinguishing between teleological and nonteleological systems is in terms of the frequency with which teleological systems approach, attain,

[2] See Beckner (1959); Canfield (1966); Hempel (1965); Munson (1971); Rudner (1966); Hugh Lehman, "Functional Explanation in Biology," *Philosophy of Science,* 32 (1965), 1–20, and William Wimsatt, "Teleology and the Logical Structure of Functional Statements," *Studies in the History and Philosophy of Science,* 3 (1972), 1–80.

or maintain certain preferred states. On this analysis, a necessary condition for a system's being goal-directed is that a certain state of the system, the goal-state, occur with a sufficiently high frequency relative to the other possible states of the system. In order to make this comparison, the elements of the system must be identified, the relevant characteristics of these elements specified, and a set of laws provided from which the interrelations of the elements or successive states of the system can be inferred. For example, the elements of the solar system are the sun and the nine planets. Each of these bodies is describable by indefinitely many properties that could function as state-variables of the system. Of these, the two that are relevant to celestial mechanics are location and momentum. A state-description of the solar system at a given time would consist in ten pairs of sentences specifying the location and momentum of each of the solar bodies at that time. The laws would be those of celestial mechanics.

The laws of celestial mechanics specify allowable states of the system. The question is whether or not a state (or set of states) of the solar system can be discovered that occurs with a frequency sufficient to entitle it to be termed a goal-state of the system. If one were to choose as the goal-state the occurrence of all the solar bodies falling in a single straight line, such a goal-state would occur very infrequently, and the solar system would not be teleological. If one were to choose the occurrence of any two planets lining up in a straight line with the sun as the goal-state, then such a goal-state would occur much more frequently, but not frequently enough for the solar system to be viewed as a teleological system. The only characterization of a goal-state for the solar system that holds out any hope of recurring with sufficient frequency for the solar system to be considered teleological is any configuration compatible with the laws of celestial mechanics. By this definition, the solar system would *always* be in a goal-state.

Another way to distinguish between teleological and nonteleological systems concerns their structure, the existence of mechanisms to promote causal feedback loops. As we observed earlier, causal sequences are better described as trees or networks than as chains. Certain of these causal networks contain closed loops; that is, an instance of a type of event *A* produces several effects, including an instance of type *B,* which has several effects including an instance of *C,* which produces several effects, including an *altered* instance of an event of type *A*. In other words, part of the output of the process is recycled as an input. In positive feedback, *A* is in some sense increased; in negative feedback, it is decreased. In ordinary discourse, people tend to call examples of positive feedback "vicious circles" if the end result is undesirable. For example, a certain girl is found unattractive by boys her own age because she is overweight. This rejection in turn makes her unhappy, which causes her to eat more, which makes her fatter, which leads boys to find her even less attractive, and so on. Examples of negative feedback are just as common. A teacher encourages a shy student to speak up until he begins to monopolize the discussion, whereupon the teacher begins to discourage his participation somewhat. The operon model discussed in Chapter One is an excellent example of negative feedback.

On the surface, the notion of negative feedback seems admirably clear, but it turns out to be extremely difficult to define in a way that distinguishes it from ordinary causal networks. For example, Nagel's definition of teleological system (in Munson, 1971) in terms of negative feedback depends on a prior knowledge of what counts as a goal-state. It runs roughly as follows:

> A system *S* is goal-directed with respect to a goal *G* during the interval *T,* if and only if, during *T* any primary variation either in *S* or in the environment *E,* within a certain range, is accompanied by adaptive variations.

The operative terms in this definition are "primary variation" and "adaptive variation." Nagel begins by defining a goal-state. A system is in a *G*-state at a particular time, if and only if the system either has the property *G,* or else a sequence of changes will take place in *S* in consequence of which *S* will possess *G* at some subsequent time. For example, either the temperature is 72°F or else the furnace is running and the temperature will reach 72°F shortly. A primary variation is a change in one or more of the state-variables of the system which would take the system out of the *G*-state if unaccompanied by changes in the other state-variables of the system. If the oven is turned on in the kitchen, the temperature of the house will rise too high unless the thermostat turns on the air-cooling unit. These variations which follow upon primary variations and compensate for them to bring the system back to its *G*-state are termed adaptive variations. A similar story could be told for the torpedo schema.

PREFERRED STATES AND NEGATIVE FEEDBACK—OBJECTIONS

In the preceding section we sketched two criteria which have been suggested to distinguish teleological from nonteleological systems. Both are relevant. Neither, however, seems to be either necessary or sufficient. The attainment or maintenance of a certain preferred state is not necessary for a system's being teleological. In certain instances, a system that is clearly teleological never maintains its preferred state; as in the case of the temperature-control schema, it oscillates around this preferred state. In other cases, the preferred state is never attained, let alone maintained. Instead it is approached periodically or asymptotically. For example, no species is ever perfectly adapted to its environment, though some approach this ideal more closely and more frequently than others. On occasion, a teleological system attains its goal, but only once. Finally, a teleological system can be teleological without being perfect. A system designed to attain a goal can consistently fail to do so because of a slight fault in its organization, a situation termed goal failure. A homing torpedo would still be a teleological system even though a defect in its homing device resulted in its always passing just behind the target ship.

Only the last of these objections to the preferred-state analysis of teleological systems is serious. This analysis could be expanded to include approach or oscillation around a goal-state as well as its maintenance or attainment. Similarly, the frequency necessary for terming a system teleological could be salvaged in cases such as the homing torpedo by treating them as classes. Each torpedo attains its goal only once, if at all, but such homing torpedos as a class tend to hit their targets more frequently than they would if they were just shot off at random or aimed at the target ship in the old-fashioned way. The example of consistent goal failure, however, indicates that something is being left out of the preferred-state analysis. A blue baby is still a teleological system even though it is nonviable. Our ability to decide that a system that is not attaining, maintaining, oscillating around, or even approaching its goal-state is teleological indicates that there is something more to teleological systems than the definition in terms of preferred states would suggest.

The chief fault of the definition of teleological systems in terms of goal-states is that it omits any reference to the mechanisms by which they are brought about. The definition in terms of causal feedback loops is an attempt to specify the general nature of such mechanisms. The two types of teleological systems that appear to be least adaptable to the negative-feedback analysis are the use of artifacts by agents when these artifacts are not themselves directively organized and systems that are directively organized but not by means of negative feedback. A hammer is made to perform certain functions (for example, driving nails, pulling nails, and so forth). It can be used to perform additional tasks (holding open a window, committing murder); it cannot be used for others (as a contraceptive device, to section an embryo). The physical characteristics of the object determine what it can and cannot be used for, but within these limits, what it is and is not used for is determined by the intentions of the agent. If the use of nondirectively organized artifacts is to be assimilated to the negative-feedback model, such a transient situation as an agent using a hammer to drive a nail must be considered as being a directively organized system. If we are to avoid reference to mentalistic entities, the directively organized character of these systems must be supplied by the organization of the agent's nervous system.

Certain preferred states are brought about in teleological systems by mechanisms which at least initially do not seem to be examples of negative feedback; for example, the utilization of alternative pathways. If the kidney cannot rid the body of excess water, sweating can compensate to some degree.

The net effect of such a system is the maintenance of a particular preferred state or series of preferred states, but such systems fail to accord either with the letter or the spirit of the usual definitions of negative feedback. Hence, the definition is somewhat too restrictive. It is also much too inclusive, applying as it does to numerous types of systems which at least initially do not seem to be teleological—a compound pendulum in a state of rest, an elastic solid, a steady electric current flowing through a conductor, and a chemical system in thermodynamic equilibrium. The negative feedback model is designed to accommodate not only those cases where a system returns to its preferred state but also to those cases in which the system progresses through a prescribed series of preferred states. In the latter cases, it fails to mention the causes of such regular sequences—the program.

The usual procedure at this point for those attempting to define the notion of teleological systems in this manner is to introduce clauses and restrictions to their original definitions in order to accommodate these and other counter examples. Some of these amended definitions come fairly close to marking out the intuitive notion of teleological systems. For example, neither the preferred-state criterion nor the negative feedback analysis seems adequate by itself to make the desired cut with sufficient precision and clarity, but taken together they might do the trick. Some example which might be an intractable borderline case on one criterion might be handled with comparative ease on the other. If a particular state of the system is clearly a goal-state, then whatever series of causal mechanisms bring it about must count as instances of negative feedback. If a particular causal sequence is clearly a case of negative feedback, then whatever state it brings about must count as a goal-state. But even if such a maneuver were successful, what is the point of such exercises? One still has to show that such an analysis is a contribution to philosophy of science and not just an exercise in linguistic analysis. In order for explications like those set out above to be philosophy of biology, some connection must be shown between these explications and current biological theory. The strength of the second pair of criteria that have been suggested to define the notion of teleological systems is that they are founded in and contribute to current biological theories.

PROGRAMS AND SELECTION PROCESSES

Neither of the preceding analyses does full justice to the notion of teleological systems, whether in general or in some restricted sense. Both criteria are too strong, excluding systems that at least appear to be teleological. They are also too weak, including systems that are clearly not teleological. Perhaps the borderline between teleological systems and nonteleological systems is hazy, but the haze does not extend all the way to the solar system and a compound pendulum at rest. Of greater importance, however, is the vagueness that surrounds the criteria themselves. How can significant goal-states be distinguished from other states of the system? How can negative feedback loops be distinguished from causal networks in general? The discouraging feature of both analyses is that they provide no basis for further analysis. Why exclude elastic solids from the class of teleological systems? Why include a migrating butterfly?

The chief advantage of the second pair of criteria which have been suggested for defining teleological systems is that they are grounded in scientific theory, chiefly evolutionary theory, learning theory, and information theory. It should occasion no surprise that evolutionary theory should play a central role in any new analysis of teleology since it was evolutionary theory that administered the *coup de grâce* to nineteenth-century teleology. The roles of learning theory and information theory are more equivocal. Theories of how organisms learn are still very rudimentary; any analysis of teleology that depends on them thus can be no better than they are. And information theory is not an empirical theory at all, but a formal theory, a branch of mathematics. At best it can serve as a formal analogy, linking the mechanisms of evolution and learning.

Ernst Mayr (in Blackburn, 1966, and Munson, 1971) has opted for a structural definition of teleological systems in terms of the presence of programs. Mayr argues that an "individual who—to use the language of the computer—has been 'programmed' can act purposively" and concludes that it "would seem useful to restrict the term teleonomic rigidly to systems operating on the basis of a program, a code of information." In the functioning of living organisms as such, the necessary information is coded into the DNA (or RNA) of the organism and can be passed on from generation to generation. In cases of purposive behavior, it is the nervous system (or an analogous system) that is programmed. For better or for worse, the information stored in the nervous system cannot be passed on directly from one generation to the next but must be acquired anew with each generation. The functioning of artifacts is assimilated to this paradigm, either by the presence of a program within them (for example, a bimetal bar in a thermostat) or by reference to the program of its manufacturer or user.

Mayr's problem, of course, is defining the notion of a program. As the quotation marks indicate in the passage cited, Mayr is aware that he is operating on the force of an analogy which is justified to some extent by the broad success of the language of information theory, but something more than an analogy is needed before his analysis can be considered fully satisfactory. Mayr is currently working on providing a more adequate definition of "program." He now also feels that the notion of teleology can best be explicated in terms of processes, not systems. However, for our purposes, the importance of his analysis is not his reliance on programs to distinguish teleological from nonteleological systems (or processes), but the explicit recourse he makes to selection theories.

William Wimsatt has opted for an historical definition of teleological systems. He argues that a necessary requirement for a system to be teleological is that it arise, either directly or indirectly, through a selection process. In a selection process, a mechanism exists for differentiating between items, for retaining some and rejecting others, and for the incorporation of the selected items into the makeup of the system. Just as organisms develop through a long process of replication, mutation, and natural selection, purposeful agents come to behave purposefully by means of the learning process, a process of selective reception and retention of stimuli as well as selective retrieval of the impulses retained. (According to one hypothesis, DNA plays a crucial role in the chemistry of learning. Supposedly memory results from the establishment of minute, closed-loop systems of DNA, RNA, and specific proteins.)

Wimsatt's analysis is confronted by two problems—first, the presentation of an adequate analysis of selection processes, and second, the need to justify his interpretation of "teleological systems" as an historical concept. At the very least, the adoption of his analysis requires a modification of current usage, because currently teleological systems are identified without any necessary reference to their origins. In point of fact, all those systems we view as being teleological arise, directly or indirectly, through selection processes. Living organisms arise through natural selection, and conscious agents become conscious through the selective retention and organization of stimuli. Few if any artifacts arise directly by means of selection processes, but they are produced by entities that did. Hence, they can be viewed as arising "indirectly" through a selection process.[3]

TELEOLOGY AND SELECTION THEORIES

All four of the criteria suggested for defining teleological systems seem in one way or another relevant. Each is somewhat vague, but these areas of vagueness tend to cancel each other out when all four criteria are used conjointly. For our purposes, however, the important feature of this cluster of defining characteristics is their eventual reliance on selection theories. For teleological systems, unlike other systems, there exists a series of laws which determine ever-decreasing preferred states of the system. To use the terminology introduced by Mayr (in Blackburn, 1966, and Munson, 1971), series of proximate and ultimate causes exist for teleological systems. Laws concerning the most proximate causes limit the possible states of the system only slightly. Given

[3] Conversations with both William Wimsatt and Mary B. Williams have helped to clarify my understanding of teleological systems.

just these laws, the system could exist in a wide range of states. It does not, because of the existence of more ultimate laws limiting its permissible states. For living organisms, the most ultimate laws are those of evolutionary theory. They place the greatest constraints on the possible states of the system. To put it differently, they determine the ultimate goal-states of the system.

For example, physiology treats the broad requirements of living organisms. Many more kinds of organisms could exist than do, and certain features of existing organisms are inexplicable just on the basis of the laws of physiology. The more ultimate laws of evolutionary theory not only explain the selection of these preferred systems of organization but also account for their physiological peculiarities. Prior to the acceptance of evolutionary theory, the chief end attributed to structures and behaviors was the maintenance of the individual. From this, the maintenance of the species was thought to follow automatically. After the acceptance of evolutionary theory, the individual organism had to be recognized as a compromise between partially conflicting goals—the good of the individual and the good of the species.

At first sight, the existence of attributes that promote the survival of the species at times at the expense of the individual organism seems surprising. It results from the inability of the genome to undergo massive reconstructions in the space of a single generation and the fact that each individual is the result of its ancestors having been able to reproduce. The individual has to survive, but it is also the result of past compromises. The possession of mammary glands by a particular mammal is not necessary to its survival; it may even be detrimental. But for this mammal to have come into existence, it is necessary for its mother to have possessed mammary glands. Any genetic mechanism that removed the genes that produce mammary glands might well benefit that particular individual, but it would also guarantee the extinction of that lineage and the loss of that mechanism. Given the inertia built into the mechanisms of genetic transmission, each individual tends to possess not only those attributes that will contribute to its survival but also those that led to its own generation. In turn it will pass on these latter attributes even though their possession may decrease somewhat the survival potential of its offspring. The cost of reproduction is high, sometimes leading to death, but it is a price all living creatures must pay. This same genetic inertia is also responsible for the continued inheritance of vestigial organs—organs that serve no positive function in the organism that possesses them but that did in some ancestral form.

In this section we have argued that the prevalence of significant, narrowly defined goal-states, brought about by causal feedback loops, especially negative feedback, governed either directly or indirectly by some sort of program, are important features of teleological systems. But of greater importance is the existence of a hierarchy of increasingly more restrictive laws which govern these systems. Teleological systems do arise, either directly or indirectly, through selection processes, but the importance of selection processes for our analysis is that the ultimate theories that determine the narrowest goal-states are selection theories. In cases of nonteleological systems, no such hierarchy of laws exist. For example, only one set of laws governs the configuration of the solar system, the laws of celestial mechanics. Simple geometric considerations are useless in defining preferred states for the solar system. One configuration is as likely to turn up as another. In teleological systems, a selection theory permits the delimitation of certain preferred states of the system, these preferred states can be defined quite narrowly, and the laws used to infer them are at least partially independent of the less restrictive laws used to infer more general states of the system.

TELEOLOGICAL STATEMENTS AND THE COVERING-LAW MODEL OF EXPLANATION

Almost all of the preceding discussion has concerned teleological systems. We now must turn to an examination of teleological statements and teleological explanations. A good strategy in investigating the relation between teleological and nonteleological language would be to contrast teleological statements with ordinary causal statements. Both can occur in discourse of roughly the same level of scientific sophistication, and the distinction marked by their use accords roughly with the felt difference that subsequent analysis is intended to explicate. Next, one might move on to compare the use of teleological and nonteleological statements in informal explanations. Finally, a more formal analysis of both teleological statements and teleological explanations might be provided and contrasted with a similar analysis of causal statements and causal explanations. Instead, the usual practice has been to equate teleological statements with teleological explanations and then proceed to judge these explanation sketches according to the standards of the covering-law model of explanation. For example, both Nagel (in Munson, 1971) and Hempel (1965) view teleological statements as elliptical arguments. Nagel thinks that teleological statements can be converted into arguments which succeed in meeting the requirements of the covering-law model by the simple expediency of equating functions and goals with necessary conditions. Hempel believes that such a maneuver will not work. He concludes that the information typically provided in a functional explanation affords neither deductively nor inductively adequate grounds for explaining the presence of the structure in question.

The mistake these philosophers are making is not the attempt to uncover the implicit content of teleological statements or teleological explanations, nor is it the tacit assumption that these formulations must necessarily be filled out in the direction of the covering-law model. Rather it is viewing every discrepancy between teleological explanations and the covering-law model as stemming from the teleological formulation's being teleological. But too many of the ways in which teleological explanations differ from the covering-law model have nothing to do with the fact that these explanations are teleological. Ordinary causal explanations differ in many of these same respects from the covering-law model. "What causes mammals to sweat?" One answer is an increase in the temperature of their environment. "What is the function of sweating in mammals?" One answer is the maintenance of a constant body temperature. Both of the questions are legitimate and the answers correct as far as they go, and in many contexts there is no reason to go any further. But from the point of view of the covering-law model, these answers are explanation sketches that need filling out. However, the filling-out process is quite difficult in both cases, involving many of the same difficulties, and what is more, frequently beside the point.

One easy way of eliminating teleological statements is to identify functions, goals, and purposes with effects, then to equate effects with necessary conditions and causes with sufficient conditions. If one accepts these identifications, then the following statements all make the same assertion:

1. The function of A is B.
2. The effect of A is B.
3. The cause of B is A.

Objections have been raised to all three of the identifications stated at the beginning of this paragraph. We will start with the objections that have been raised to the identification of effects with necessary conditions and causes with sufficient conditions. It should be noticed at the outset that these objections have nothing to do with teleological discourse as teleological discourse. Rather they stem from the existence of multiple causes and multiple effects, a situation not peculiar to teleological systems. Next we will turn to the objections raised to the identification of functions with effects. These objections have their source in the notion of a positive function.

No one discusses teleological statements without noting the existence of functional equivalents, alternative ways in which the same function can be performed. There are too many ways for sap to rise in trees, rabbits to ovulate, and the same phenotypic character to be produced. Hence, functions cannot be interpreted simply as necessary conditions. But this conclusion follows from the existence of multiple causes and is not peculiar to teleological systems. Several authors have also noted the existence of goal-failure, the failure of a teleological system to attain its end. A function of the heart in mammals is to circulate the blood even when it is not doing so in the body of a blue baby. A prizefighter can practice diligently to win a fight he eventually loses. In short, teleological systems can be teleological without being perfect. But this condition is just as prevalent in causal discourse. Just as the existence of functional equivalences entails that a particular structure or behavior is not necessary for the performance of a particular function, the existence of goal-failure entails that none is sufficient either. Just as death can be produced by a host of alternative causes, investing in the stock market can have just as many alternative effects. On this score, causal and teleological statements are in the same boat.

Ernest Nagel (in Munson, 1971) has suggested that the difference between teleological and nonteleological statements is "comparable to the difference between saying that Y is an effect of X and saying that X is a cause or condition of Y. In brief, the difference is one of selective attention." If we keep in mind the existence of multiple causes and multiple effects, there is some truth in what Nagel says. Teleological statements do not require any temporal inversion of causal sequences. The superficial appearance of temporal inversion stems from selective attention. In ordinary causal statements, we direct our attention to an earlier event that has caused the event in question, whereas in a teleological statement we direct our attention to a later event to see what the event in question causes. It makes no difference whether we say that the maintenance of a constant temperature is an effect of sweating in mammals or that sweating in mammals is a cause of the maintenance of a constant temperature, but neither of these statements seems equivalent to the teleological claim that a function of sweating in mammals is the maintenance of a constant temperature. The identification of functions with effects seems illicit. All functions are effects, but not all effects are functions. After all, dehydration is also an effect of sweating, yet dehydration is not a function of sweating. Similarly, the heartbeat in mammals not only circulates the blood but also produces heart sounds. The production of heart sounds is hardly a function of the heartbeat in mammals.

The resolution of this apparent difference between functions and effects is quite straightforward. We must distinguish in teleological systems between "function" used in a nonevaluative generic sense and "function" used in the sense of a positive function. In its generic sense, "function" is equivalent to "effect"—as long as "effect" is not identified with "necessary condition." In this nonevaluative sense, dehydration is a function (an effect) of sweating and the production of heart sounds is a function (an effect) of the heartbeat. Usually, however, "function" carries a positive connotation, implying that the structure or behavior contributes to the approach, attainment, or maintenance of a preferred state of the system. In this context, a structure or behavior can also be dysfunctional or nonfunctional. A structure is dysfunctional if it impedes the approach, attainment, or maintenance of the preferred state of its system. A structure or behavior is nonfunctional if it neither helps nor hinders. The existence of nonfunctional structures or behaviors is solely a consequence of how finely we distinguish positive functions. The finer the evaluative net, the fewer structures or behaviors that count as nonfunctional.

TELEOLOGICAL EXPLANATIONS AND THE COVERING-LAW MODEL OF EXPLANATION

Teleological explanations contain at least one operative teleological statement. Causal explanations are given in terms of causes. Teleological explanations are given in terms of functions, goals, and the like. In contrasting teleological explanations with the covering-law model, three questions arise: Can teleological explanations be recast as valid deductive arguments? If not, can they be recast as warranted inductive arguments? In either case, do the operative teleological statements in the explanations count as genuine scientific laws?

Suppose someone wanted to know why plants contain chlorophyll. One might reply, "To serve as a catalyst in photosynthesis." But what is the function of photosynthesis? To produce the necessary nutriments for the survival of the plant. If these functional statements are taken as expressing necessary connections, this teleological explanation can be reformulated as follows:

1. In plants, chlorophyll is necessary for photosynthesis.
2. Photosynthesis is necessary for survival in plants.
3. Plants do exist.

Hence, photosynthesis is being carried on in plants.
Hence, chlorophyll exists in plants.

This example, like so many examples from the philosophical literature, needs a little amending, because the first two premises are false as they stand. Chlorophyll is not necessary for photosynthesis—other closely related compounds also can serve as catalysts. Nor do all plants contain chlorophyll or these other related compounds. Some plants are parasitic, obtaining their nourishment from the living tissues of their hosts; others are saprophytic, living off dead organic matter. Furthermore, chlorophyll is not always necessary for survival even in those plants that possess it, because in the absence of light some of them can become parasitic or saprophytic. To make matters a little more complex, some animals also possess chlorophyll.

The problem is, of course, the prevalence in teleological systems of functional equivalents—several items that can serve the same function, or several behaviors that can accomplish the same end. Given the current grouping of organisms into taxa and analysis of organisms into functional parts, any function in a given group of organisms can be performed in more than one way. Two alternative strategies have been suggested to eliminate functional equivalents and hence salvage the deductive character of teleological explanations. One way is to define the parts of teleological systems functionally. For example, Beckner[4] has suggested that the pumping of blood should be a necessary condition for a structure being a heart. In our example, the possession of chlorophyll would be a necessary condition for an organism being a plant. Hence, mistletoe would no longer count as a plant.

Two objections can be raised to Beckner's suggestion. First, it is strongly opposed to current scientific usage. When a hammer is used as a doorstop, does it become a doorstop? When a button is used as a pawn, does it become a pawn? In contexts such as these, such questions seem all but unanswerable, but in the case of science, there are good reasons for preferring one answer to another. Chemical reactions currently are not defined in terms of a specific catalyst. Usually several will do. Nor are the names of taxa defined in terms of the possession of a single trait such as the presence of chlorophyll. No trait is strictly necessary for taxon membership. Sometimes there are good reasons for introducing stipulative definitions into science. In this case, however, the only reason for legislating functional definitions for such terms as "photosynthesis," "plant," and "heart" is to salvage the deductive character of functional explanations. But even if one were inclined to pay such a price, the maneuver fails, because teleological statements cannot count as scientific laws under such an interpretation. Instead they become logical truths. If hearts are defined in terms of pumping blood, then the claim that hearts pump blood ceases to be an empirical statement and teleological explanations cease to be scientific explanations.

A second strategy for making teleological explanations deductive is to reanalyze teleological systems to eliminate functional equivalents. Given the particular analysis, it will then be a contingent fact that a particular item or behavior is necessary for a particular function. The situation is similar to that discussed earlier in connection with causation. If causal laws are to be process laws, then there must be universal covariation between causes and effects. If phenomena are analyzed into causal trees and causal networks, as is currently done in molecular biology, then universal covariation is impossible. There are always multiple causes and multiple effects. For causal trees and networks, it may be possible to reanalyze the situation to obtain the requisite universal covariation—it has been done successfully at times. However, there are good reasons for thinking that this strategy will not work in biology as long as living organisms, biological species, and the like are considered the relevant systems. As was pointed out earlier, the commonest mechanism for promoting adaptability is redundancy—several ways of performing the same function. Hence, living systems frequently contain redun-

4 "Metaphysical Presuppositions and the Description of Biological Systems," in J.R. Gregg and F.T.C. Harris, *Form and Strategy in Science* (Dordrecht: D. Reidel Publishing Co., 1964).

dancies, or what we have been calling functional equivalents. If the phenomena under investigation are to be reanalyzed to eliminate functional equivalents, living organisms cannot be treated as systems in this analysis.

Teleological systems do seem to force themselves on us. If anything is to count as an individual, surely a living organism must. Biological species seem almost as apparent—aborigines recognize them as readily as trained naturalists. Even though teleological systems are largely open, they somehow remain recognizably the same through continual change. There is a permanence about teleological systems, but it is a permanence unlike that of the pyramids which stay the same by changing as little as possible. Teleological systems persist by changing. Scientists have been most successful in dealing with effectively closed systems like the solar system. Because these systems are effectively closed, process laws can be formulated governing their behavior. At first teleological systems appear to be more completely closed than they actually are because, like closed systems, they maintain their integrity. Thus, scientists are lured into thinking that they can discover universal process laws governing their behavior. But the maintenance of this integrity is accomplished by means of a constant inflow and outflow of energy, through numerous alternative pathways. If it were not for this retention of pattern, it would never occur to a scientist that such a complex could be treated as a unit.

The issue, however, is not how apparent conventional biological units are to human beings, given our relative size and the peculiarities of our sense organs. Organisms and biological species tend to be apparent to human beings, but their status as important biological concepts would be unchanged if they were as obscure as gravitational fields. The issue is whether they can contribute to the formation of improved biological theories. Perhaps biological species do not form significant evolutionary units—perhaps the emphasis in the past on such units was mistaken. Future biologists may produce new biological theories, perhaps new versions of evolutionary theory, that make no reference to organisms or the reproductive relations between them. Perhaps, but it should be kept in mind that the conceptual revolution necessary for such a change would be enormous—so enormous that one would need a more compelling reason for undertaking the task than the desire to eliminate functional equivalents and, thus, make teleological explanations deductive.

A third strategy for making teleological explanations deductive is to list all the functional equivalents for a particular function and deduce the extremely weak conclusion that at least one of the items listed was responsible for performing it. For example, plants can fulfill their nutritive requirements parasitically, saprophytically, or by photosynthesis. Hence, if a plant is continuing to exist, one can deduce that it is doing so by at least one of these methods. The deductive character of the argument has been salvaged, but at the expense of weakening the conclusion considerably. One way of strengthening the conclusion is by eliminating the functional equivalents one at a time in the manner suggested by Francis Bacon in Elizabethan England and John Stuart Mill in Darwin's day. In order for either of these strategies to work, there must be a finite number of sharply distinguishable alternatives for each function. Rarely, in actual examples, does this seem possible. The existence of functional equivalents is a fact of nature that no amount of additional knowledge will eliminate.

One last strategy still remains—to interpret teleological explanations as examples of statistical-probabilistic explanations. Hempel (1965: 312) addresses himself to this possibility. He says that perhaps a teleological explanation "could more adequately be construed as an inductive argument which exhibits the occurrence of [an item] as highly probable under the circumstances described in the premises." But he continues, "This course is hardly promising, for in most, if not all, concrete cases it would be impossible to specify with any precision the range of alternative behavior patterns, institutions, customs, or the like that would suffice to meet a given functional prerequisite or need. And even if the range could be characterized, there is no satisfactory method in sight for dividing it into some finite number of cases and assigning a probability to each of these."

The examples Hempel examines in his paper are from functional analyses in sociology. In biology, perhaps one cannot list all the functional equivalents for a particular need and assign a probability to each, but quite frequently the most probable alternatives are few and well known. Given evolutionary theory, it is highly likely that any structure in an organism will fulfill a positive function and will do so with a high degree of efficiency. There are exceptions, of course, but they are relatively rare. In almost every example given thus far of teleological statements in biology, the item or behavior mentioned performs its function in almost every instance, even in extremely abnormal circumstances. Normally a positive function of the heartbeat in mammals is to circulate the blood, and in most instances that is precisely what it does. In teleological systems, there are almost always functional equivalents, but usually not many that come into play very often, and the most likely ones are well known. Of course, not all teleological explanations are warranted. Sometimes there are too many functional equivalents and none of them is any more prevalent than any of the others. In such cases, it is impossible to infer which functional equivalent is performing the function in any one instance with any respectable degree of probability. But such situations seem to be the exception and not the rule.

However, even if teleological explanations are interpreted on the pattern of statistical-probabilistic explanations, a major problem still remains. The generalizations in statistical-probabilistic explanations must be scientific laws. In teleological explanations, these generalizations are teleological statements. Hence, teleological statements must count as scientific laws. Unfortunately, the examples given thus far do not look propitious. For instance, even if it were true that chlorophyll is necessary for photosynthesis, this hardly seems a law of nature. Usually catalysts are not necessary for reactions—all they do is increase efficiency, and in most cases, widely different catalysts will do the job. By all appearances, the statement that chlorophyll (or a closely related compound) is necessary for photosynthesis is an accidental generalization, not a law of nature. Many philosophers and scientists have been dissatisfied with teleological explanations in the past, but the locus of their dissatisfaction has been misplaced. The real problem concerns the *lawfulness* of teleological statements. Perhaps not all of them are natural laws, but at least some of them must be if teleological explanations are to count as covering-law explanations. (See earlier discussion of laws and accidental generalizations in Chapter Three.)

SUMMARY AND CONCLUSION

Philosophers have often remarked that causal statements and explanations are made against the background of a theory. This theory determines what items are to be considered elements in the causal network and how they are to be related. But similar observations can be made about teleological statements and explanations. They too are made against the background of a theory. Theories determine what is to count as the system, a part of the system, and the preferred states of the system. But in the case of teleological statements and explanations, these theories are selection theories. There is a tendency on the part of philosophers to assume that causal discourse is inherently more informative than teleological discourse. It is always teleological discourse that must be reformulated in purely causal terms. It is certainly true that teleological claims can be quite casual, made on the basis of extremely rudimentary theories—so rudimentary that they hardly deserve to be considered theories at all. This is typically true of teleological claims made in ordinary discourse, and that is why nothing very precise can be said about teleological statements and explanations in ordinary discourse. For example, most purposive claims are made against nothing more than common sense theories of the "person." But similar observations can be made about ordinary causal discourse as well.

Requests for explanations are sometimes quite casual and call for equally casual explanations, but science is not a casual affair. Both causal and teleological explanations presuppose considerable background knowledge, and the assumption is that it can be produced on demand. Too often the requisite background knowledge is all but absent and the apparent explanatory power of the formulation is specious. One of the dangers of both teleological and causal explanations is that often they sound more explanatory than they actually are, and further inquiry is inhibited. However, the danger in the case of teleological explanations seems to be greater than that for causal explanations because teleological systems seem so incredibly apparent to us. For example, G. C. Williams (1966: 260–61) admits that in his discussion of teleonomy, he has assumed "as is customary, that functional design is something that can be intuitively comprehended by an investigator and convincingly communicated to others. Although this may often be true, I suspect that progress in teleonomy will soon demand a standardization of criteria for demonstrating adaptation, and a formal terminology for its description." The main reason biologists have not done so earlier is that so many of the problems that arise in determining functional relationships can be "so readily solved intuitively."

Because teleological systems seem to be far more closed than they actually are, we are led to believe that their organization and operation are far less problematic than they actually are. How sophisticated a theory do we need to recognize an organism as a teleological system? The answer is, not very, and that is why teleological explanations so often prove to be unsatisfactory. The explanation for this peculiar state of affairs is probably the fact that we ourselves are teleological systems and arose through the same selection processes as these other teleological systems, in part because of our ability to recognize and to respond appropriately to them as coherent entities.

The difference between teleological and causal discourse is that each lends itself to the description of different types of partially open systems. Neither is inherently more informative than the other in all contexts. Sometimes the causal formulation is more informative and the teleological statements must be expanded; sometimes the teleological statement is more informative and its causal counterpart must be expanded. If causal and teleological statements are to be made intertranslatable, both must be extensively reconstructed. The important point to notice, however, is that difficulties in translation do not stem just from features that are peculiar to teleological discourse, but from features of causal discourse as well.

Causal and teleological statements are used to describe systems that are partially open, but the type of openness is different in the two cases. Situations that lend themselves to causal description are open because the causal net stretches out in all directions. Not all strands, nor all nodes in the net are equally important. Usually a causal ascription explicitly mentions only one strand in the net, perhaps only a single node—everything else is assumed as background knowledge. This is one reason that the specification of necessary conditions seems so explanatory, in fact more explanatory than is reflected in the notion of increase in likelihood. Although alternative pathways can lead to an event under investigation, the specification of a necessary condition at least indicates one node through which all of these alternative strands must pass.

Situations that lend themselves to teleological ascriptions are open because the organization of teleological systems is maintained by a continual exchange of energy and modifications in the parts of the system. When one attempts to redescribe teleological systems in purely causal terms, one is increasingly driven to do so by means of closed feedback loops. Of course, as science develops, all such situations may be successfully reanalyzed into closed systems governed exclusively by process laws on the model of the solar system. If so, both causal and teleological language will be eliminable from scientific discourse. But until then, both modes of description and explanation will remain part of the scientific enterprise and will have to be treated adequately, if not respectfully, by philosophers of science.[5]

[5] W. Wimsatt, "Complexity and organization," in K. Schaffner and R. C. Cohen, *Philosophy of Science Association, 1972,* forthcoming.

CHAPTER FIVE

Organicism and Reductionism

THE NATURE OF THE CONTROVERSY

In a recent autobiography, Julian Huxley relates the following conversation with J. S. Haldane:

> Dr. Haldane called himself an organicist, which implied being anti-mechanist and yet not a mystic vitalist—I never quite grasped what he really meant. At any rate it led to some passages at arms. As I was describing some experiment which demanded a mechanistic explanation, he burst out with "But it's a norganism, my dear young fellow, a norganism."[1]

Unfortunately such obscure utterances are all too characteristic of the exchanges between the so-called mechanists and organicists, materialists and vitalists, reductionists and holists, to mention but a few of the terms used to characterize the two sides of this perennial dispute. Yet none of these terms can be defined with any clarity. No two people seem to use them in the same way. This terminological chaos is surpassed only by the facility with which one side cheerfully caricatures the views of the other and then howls in indignant outrage because its own views have been misconstrued.

[1] *Memories* (New York: Harper and Row, 1971), p. 138. Julian Huxley is the grandson of T. H. Huxley and the brother of Aldous Huxley; J. S. Haldane was the father of J. B. S. Haldane, a founder of the genetical theory of evolution.

If it were at all possible, I would avoid entering into this dispute, but it has played too central a role in the philosophy of science for me to ignore totally. In all honesty, however, I must admit that I do not do so with any enthusiasm. It seems as if the only acceptable alternatives in this altercation are uncritical acceptance or outright rejection, and I am inclined to neither. To make matters worse, as soon as one has voiced the right slogans and termed oneself an organicist or a reductionist, denunciation is guaranteed from the opposition and at least polite silence from one's nominal allies. The tenor of the controversy is more reminiscent of political polemics and biblical exegesis than science. Of course, scientific disputes are not as different from those in politics and religion as propagandists for the scientific community would have us believe, especially when these differences concern theoretical commitments and fundamental metaphysical beliefs. But at least on some issues, debates in science rely more heavily on evidence, argument and reason than do comparable disagreements in areas like politics and religion. The question before us is the extent to which the organicism-reductionism controversy rests on differences in empirical fact, theoretical commitment, philosophy, or metaphysics. And given an answer to this question, how can it be resolved?

During certain stages in its history, the debate was carried on as if it could be settled by recourse to empirical facts. Experiments were performed, tests run, and organic substances synthesized, but to no avail. At other stages, the issues seemed to be largely theoretical. For example, mechanism was commonly interpreted as the view that all science would ultimately be reduced to mechanics, a special branch of physics. But mechanics proved to be inadequate as a basis even for physics, let alone all of science. One would think that such a development would have dealt a serious blow to the mechanistic program. Not in the least. Conversely, one would have thought that the introduction of evolutionary theory in the nineteenth century and the discovery of the molecular basis of heredity in the twentieth would have had a significant impact on the debate between the mechanists and the organicists. To the contrary. On the one hand, such antireductionists as Marjorie Grene and Michael Polanyi denounce the synthetic theory of evolution as being a reductivist dogma, treating organisms as if they were just so many heaps of molecules. On the other hand, such organicists as G. G. Simpson and Ernst Mayr view evolutionary theory as the major barrier to mechanistic reductionism. One does not have to study the history of this dispute too extensively to become convinced of at least one thing—nothing about the empirical world could possibly settle it one way or the other.

Hilde Hein (in Hintikka, 1969:239) suggests that the persistence of this controversy, the ill-humor in which it is conducted, and the ease with which any development in science can be bent to support either side indicate that it hinges on largely metascientific or metatheoretic considerations. Michael

Simon (1971:237) agrees but replies that the "clash between opposing commitments is the hallmark, not the antithesis, of a genuine scientific controversy." According to one currently popular view, the scientific enterprise is no more a matter of reason, argument, and evidence than are political revolutions. Perhaps so, but important differences exist between scientific and political controversies, on the one hand, and the vitalism-mechanism dispute, on the other. On occasion at least one scientific theory prevails over another. The scientific community opts for the heliocentric system over the geocentric. Similarly, on occasion at least, political battles are settled decisively. One side wins; the other loses. But the controversy between organicism and reductionism goes on forever.

The only trend discernible in the massive literature on the vitalism-mechanism debate is that early positions were comparatively straightforward, later positions more obscure. Originally, vitalism was the view that living creatures were different *sui generis* from inanimate objects. Either they possessed a unique form or they were made out of a different kind of stuff. Life was not reducible in any sense to non-life. Mechanists maintained that the science of mechanics, narrowly defined as the science of matter in motion, was an adequate explanatory base for all of science, including biology. To the extent that positions such as these can be disproved, they have long since been shown to be crudely mistaken. Successive generations of philosophers and scientists have modified them until they now have become so sophisticated it is nearly impossible to tell them apart. In the following pages, the vicissitudes of the controversy over the nature of life and the relation between biology and the other sciences will be traced from the early simplistic versions of both sides to the more sophisticated versions now popular. At the outset the reader must be warned that metatheoretic terms like "organicism" and "reductionism" have changed their meanings during the course of this dispute just as rapidly and haphazardly as such theoretic terms as "force" and "species." Each of these terms can no more be defined and these definitions used throughout the course of the discussion than could "gene" be defined at the beginning of Chapter One and used consistently throughout the discussion of Mendelian and molecular genetics.

LIFE AS A VITAL FLUID

The most straightforward version of vitalism is that living creatures differ from inanimate objects because they are made of different substances. Living creatures are made up of one kind of substance, inanimate objects another, and neither substance is reducible to or derivable from the other. Vital substance is not made of material substance, and material substance is not made of vital substance. A slightly more sophisticated version of this type of vitalism is that everything is made of the same basic kind of substance, except that living creatures contain an additional vital substance. After Newton, vital substance was most frequently characterized as a fluid, in analogy to caloric, phlogiston, and other imponderable fluids popular in the day. Just as heat was considered to be a fluid that flowed from warm bodies into cold ones, life was considered a vital fluid that was passed on in reproduction and departed upon death.

If one accepts life as a vital fluid, two consequences necessarily follow: living creatures could not have evolved from inanimate substances, and life cannot be created by man in the laboratory. The arguments for these two positions are quite similar: if life is a vital fluid and the earth was originally devoid of this fluid, living creatures could never have arisen naturally here on earth. Similarly, a biochemist might combine all the right physical elements in the right order, but he could never create life in a test tube unless he added some vital fluid. Because the biochemist would be at a loss as to how to obtain any of this vital fluid, he could never produce a living creature. An analogous argument has been urged against the idea of a computer thinking. If mind is a special kind of substance that people have and machines lack, a scientist might produce a computer that could perform any of the mental feats that a person can perform (for example, play chess, solve mathematical problems, make inductive inferences, write poetry, be deceived, and so forth), but it could never be conscious because it would lack the requisite mental substance.

The preceding positions are extremely simplistic—so much so that it would be difficult to find a scientist today who holds any of them. One would think that the synthesis of urea in 1828 and later accomplishments, including recent successes in producing replicating segments of DNA, must have played a significant role in the abandonment of this version of vitalism. Such is not the case. Just as astronomers abandoned the Ptolemaic system long before the observation of stellar parallax in 1837, most biologists became disenchanted with the notion of vital fluid long before its existence was decisively refuted. In the main, biologists abandoned the notion because it led to no advances during the course of their researches and because other theories arose which were incompatible with it—specifically, evolutionary theory. If living creatures generated spontaneously from purely inanimate substances in the distant past, then life cannot be a vital fluid. Hence, it is at least in principle possible to create life in a test tube.

LIFE AS A VITAL FORCE

Some advocates of vitalism argued that life was not a vital fluid but a vital force. The issues here were not as straightforward, because of pervasive confusions over the ontological status of forces in the physics of the time. Some physicists argued that forces, like the force of gravity, were operative in causal situations. Gravity caused things to move about. Other physicists insisted that gravity was not a thing, but a property of material bodies. (See the dispute between William Whewell and John Stuart Mill for the two sides of this issue.) A similar confusion surrounded the notion of vital forces. If vital forces are given a substantial interpretation, then they seemed as repugnant to latter-day biologists as vital fluids and were rejected for the same reasons. However, if they are viewed as properties of material bodies, vitalism becomes a coherent position and the major objections to it can be circumvented. Just as magnetism is not some imponderable substance added to iron bars when they become magnetized, life is not some imponderable fluid added to living creatures at conception. Both magnetism and life result from the organization of the material out of which they are made. Each molecule of iron exhibits a magnetic effect, but in ordinary iron bars, these magnetic effects cancel each other out because the molecules are arranged in a haphazard way. However, as more and more molecules come to be arranged with their poles aligned in the same direction, the bar gradually becomes more and more magnetic. The situation is analogous to that in physics when physicists ceased to view time as something that flowed like a river and space as a receptacle that extended indefinitely in all directions regardless of the existence of material bodies. Both space and time are now viewed as organizational properties of material bodies. They are not fictions, unless one wishes to term all properties fictions.

There are not many trends discernible in science, but one of them seems to be the shifting of key scientific concepts from the category of things and substances to the category of properties, especially relational and organizational properties. Life is no more a thing than is time, space, gravity, or magnetism. One might well add *mind* to this list. If life is viewed as an organizational property of certain material systems, then we do not have to worry about where life came from when the first living creatures arose from inanimate substances or where it goes when a living creature dies, any more than we have to concern ourselves with discovering where magnetism comes from when an iron bar becomes magnetized or where it goes when the bar becomes demagnetized. In short, we need not postulate a magnetic heaven as a final resting place for good magnetic fields.

THE MECHANISTIC VIEW OF LIFE

Early versions of mechanism were characterized by two basic tenets: the belief that all of science could be derived from the science of mechanics and the resulting conclusion that living creatures could be treated as machines pure and simple. Physicists abandoned the first tenet by at least the turn of the century when they discovered that many kinds of purely physical phenomena could not be explained just in terms of matter in motion. In actual fact, however, it was the machine analogy that had been most operative in the mechanistic world view in biology, and it survived even though the premise from which it supposedly follows had been discarded. Mechanists maintained that if biological phenomena were to be comprehensible, living creatures had to be interpreted quite literally as machines made up of pulleys, pneumatic tubes, and the like. Organisms might contain fluids, but these fluids obeyed purely mechanical laws.

Unfortunately, much of the current polemics against both vitalism and mechanism is directed against the earliest, most primitive formulations of these positions. For example, Michael Polanyi, who certainly should know better, argues that his recognition of "a whole sequence of irreducible principles transforms the logical steps for understanding the universe of living beings. The idea, which comes to us from Galileo and Gassendi, that all manner of things must ultimately be understood in terms of matter in motion is refuted."[2] But this idea was refuted long ago and by considerations much more decisive than any of those mentioned by Polanyi. Whether or not mechanists ever believed that organisms were heaps of molecules, they no longer do, any more than contemporary organicists believe that life is a vital fluid.

Mechanists and organicists now seem to agree that living creatures differ from inanimate objects primarily in having different types of organization (see Chapter Four). Of the two issues that still divide them, one is ontological, the other methodological. The ontological issue concerns the nature of organization itself. Is the organization of a system something over and above the arrangement of the elements that make it up, as the antireductionists seem to maintain? Must additional ontological levels be introduced for life and possibly mind? As in Darwin's day, the ontological level that is of primary interest to such opponents of mechanistic reductionism is the one that separates man from all the rest of creation. It is difficult to discover how one decides where and how many additional ontological levels to introduce. Why not introduce them for the evolution of every species, for the emergence of star systems, or the development of the corporate state? Marjorie Grene[3] argues that a one-level ontology is inadequate to account for the major areas of human experience, both in and out of science. A many-leveled ontology is necessary. But she does not tell us how we are to decide on the number of levels or where to introduce them. Adolf Portmann recognizes what he calls "centricity" in plants; Grene finds such a notion beyond her comprehension. But a critic (this one included) might well reply that he is unable to comprehend the notion of centricity as such, whether in plants or animals. As obscure as these issues may be, they are equally im-

[2] "Life's Irreducible Structure," *Science* 160 (1968), 1308–12.
[3] "Biology and the Problem of Levels of Reality," *The New Scholastic,* 41 (1967), 427–49.

portant. Perhaps current attempts at solution are too vague and impressionistic to be of much use, but eventually solutions must be forthcoming if we are ever to understand living organisms thoroughly.

It is certainly true that nothing is more obvious in the study of nature than the existence of complexity and levels of organization. Nowhere are the levels of organization more stratified and the complexity more complex than in the organic world. But ontological levels, individuals, parts, wholes, and so forth are hardly the "givens" of experience—rather these notions emerge as phenomena are investigated and need not coincide with common sense notions. For example, nothing seems as obvious to the casual observer as the difference between bushes and trees, yet this apparent difference is of no consequence to contemporary botany. Similarly, at one time the difference between sublunar and superlunar phenomena was thought to be crucial. No law that applied in one region could possibly apply in the other. Now this distinction is thought to be of no consequence. The fate of the distinction between plants and animals, between man and other species, and perhaps between the races of man is yet to be decided. Man is qualitatively different from other species. But magnets are qualitatively different from other iron bars.

Of greater significance to biology are the ways in which different biologists analyze their phenomena and organize the laws and theories that result. One biologist might treat an ecosystem as his system and study the interrelations of its parts. The parts for him would be individual organisms, features of the terrain, seasonal changes in the weather, and so on. Nor would he classify organisms according to their evolutionary descent as they are in taxonomy, but according to their roles in the ecosystem. Two species might be members of the same genus, but if one were a nocturnal carnivore and the other a diurnal herbivore, they could belong to different phyla for all that would matter. Another biologist might concern himself with the gross anatomy of the members of a particular species. Muscles, arteries, and organs form the parts of his system. A physiologist would look at this same system but differently. At another level of analysis, the cell might be the system under investigation. For the purposes of such investigations, the origins of these cells might be of no consequence. Two cells might be treated as being the same even though one came from a member of one species, the other from a second, just as two carbon atoms are identical for a physicist even though one comes from a hunk of coal and the other from a diamond.

No one questions the fact that all of these phenomena are part of one and the same reality, but it is one thing to recognize this fact and quite another to construct scientific theories that reflect this presumed unity. Numerous theories exist that account for a wide variety of phenomena at various levels of analysis, but they currently are not organized into a single unified theory. Instead, considerable contingency exists between the phenomena at these various levels as they are now analyzed—and in some cases the analyses are even incompatible. One of the contentions of mechanistic reductionism is that in the course of time, scientists will produce theories that can be synthesized into a single unified theory. Many scientists, especially biologists and social scientists, argue against both the possibility and desirability of such a program, fearing that the unity of science really means the unity of physics. But presuming that such a super theory is possible, it surely will look no more like contemporary physical theories than contemporary biological theories. Well then, what shall we call it? Call it mah-jongg for all that it matters.

REDUCTIONISM AND COMPOSITIONISM

Only the latter of the preceding considerations play an important role in current disputes among *scientists* over reduction; the controversy has taken a methodological turn. G. G. Simpson (1961) contrasts two alternative approaches in science, termed *reductionism* and *compositionism.* The issue is not whether living organisms can be dissected into parts or synthesized out of inorganic material, but one of the comparative fruitfulness of alternative methods of scientific investigation and subsequent explanation. Reduction in Simpson's sense is the relating of "structure and events at one level of organization with those of lower levels." In pure reduction, actions of the elements of a system are explained in total isolation from any system at a higher level to which they might belong. Simpson is led quite naturally to conclude that "pure reductionism offers no adequate explanations and permits no sound predictions. The action of an atom, a DNA molecule, or an enzyme, can be neither explained nor predicted outside the context of the system in which it occurs." In biology, "a second kind of explanation must be added to the first or reductionist explanation in terms of physical, chemical, and mechanical principles. This second form of explanation, which can be called compositionist in contrast with reductionist, is in terms of the adaptive usefulness of structures and processes to the whole organism and to the species of which it is a part, and still further, in terms of ecological function in the communities in which the species occurs." To the reductionist and compositionist explanations one must also add historical explanations in terms of evolutionary development.

Simpson believes that the reductionist and compositionist modes of explanation are not competing methods of explanation but complementary. Pure compositionist explanations are as incomplete as pure reductionist explanations. Molecular biologists tend to emphasize one aspect of biological systems, organismic biologists another. Only the two types of explanation, in conjunction with historical explanations, provide an adequate explanation of biological phenomena. Simpson's discussion, applied to systems in general, seems admirably well-taken, but he apparently thinks that the distinctions he makes serve to distinguish biology from physics. In biology both types of explanations can and must be provided, but apparently there are physical phenomena for which reductionist explanations are the only meaningful and allowable types of explanation. It is this implication of Simpson's arguments with which I disagree (see earlier discussion of historical and teleological explanations). The *degree* to which biologists must pay attention to organization in the explanation of even the simplest vital processes may be greater than that necessary for physicists in explaining purely physical systems. The *extent* to which records of the past history of living organisms are retained in their organization, especially at the level of individual organisms and their genomes, may be greater than that to be found for inanimate objects. The *amount* of contingency involved in the explanations of biological phenomena may be greater than is characteristic of certain very restricted kinds of physical phenomena. But these differences are matters of degree and not kind. If one examines actual physical theories, and not the usual textbook examples, one discovers that the differences between biological and purely physical phenomena do not look so great. Biologists do not have a corner on organization. Graphite and diamond are extremely different gross substances. Yet they are both made out of one and the same element—carbon—and nothing else. All their gross differences stem solely from differences in the organization of their constituent molecules.

If Simpson intends to show a sharp methodological break between physics and biology, he has not succeeded. At best, he has shown that perhaps a difference in emphasis exists. The only exceptions to this conclusion are explanations that depend on the peculiar types of organization to be found in teleological systems. If it can be successfully argued that the *only* truly teleological systems are those that arise through natural selection or through the intervention of some organism which has, and that *any* system for which either of these conditions obtain are automatically and exclusively in the province of biology, then grounds have been provided for the uniqueness and autonomy of biology. So far neither of the elements in the antecedent to this conditional statement has been firmly established.

Ernst Mayr (in Mendelsohn, 1969:128) has argued for even a stronger position than that set out by Simpson, but it concerns the logic of discovery, not alternative modes of explanation.

> Finally, it has never been demonstrated that reductionism works, so to speak, upwards. To be sure, most of the phenomena of functional biology can be dissected into physical-chemical components, but I am not aware of a single biological discovery that was due to the procedure of putting components at the lower level of integration together to achieve novel insight at a higher level of integration. No molecular biologist has ever found it particularly helpful in work with elementary particles.
>
> In other words, it is futile to argue whether reductionism is wrong or right. But this one can say, that it is heuristically a very poor approach. Contrary to the

claims of its devotees, it rarely leads to new insights at higher levels of integration and is just about the worst conceivable approach to an understanding of complex systems. It is a vacuous method of explanation.

Except for his parting comment, Mayr's claims concern the fruitfulness of reduction as a method of discovery. He is contrasting "downward reduction" with "upward reduction." In downward reduction, systems at one level are physically decomposed into subsystems. Apparently such a method affords novel insights at these lower levels of integration. In upward reduction, the components of a system are physically combined to produce higher-level systems. According to Mayr, such a procedure rarely leads to new insights at higher levels of integration. Hence, it would seem that neither method helps much in understanding systems at the level at which they exist.

It is difficult to apply Mayr's distinctions to actual examples in biology. Mayr himself has divided populations of organisms into subpopulations, recombined them in various ways, and studied the outcomes. His conclusions about the nature of the evolutionary and genetic processes were applied to species as such, and not just to the subpopulations under investigation. Similarly, one of the most fruitful lines of research in cytology involves growing specialized cells in carefully controlled conditions outside the body. The closest example of the type of reduction which Mayr seems to have in mind can be found in the work of S. Ochoa and his colleagues, who spent years building up segments of DNA, nucleotide by nucleotide. Segments have recently been constructed that are large enough to replicate. Perhaps this strategy has not been as productive as others, but certainly such investigations helped crack the genetic code, and this knowledge has had ramifications for higher levels of integration. For example, many organic diseases are caused by the alteration of a single codon.

At any one level of analysis, there is always a residue of unsolved problems. One excellent strategy in science (and not just biology) is to see if some of these residual problems can be solved by moving to a different level of analysis, sometimes higher, sometimes lower. Anyone studying the microstructure of the mammalian kidney would remain forever puzzled by certain features of the renal corpuscle if he refused to raise his vision to the level of evolving species. Conversely, a comparative anatomist would fail to recognize how serious a problem is presented by Henle's loop in the kidney if he were ignorant of cellular physiology. If mechanistic reductionism is interpreted as the claim that all problems and all solutions are legitimate only at the lowest level of analysis (i.e., the level of quantum theory), then it is surely mistaken. All scientists (and not just biologists) use both the method of analysis and the method of synthesis to investigate the phenomena that interest them. Perhaps the need for the method of synthesis is more apparent in biology because the existence of highly organized systems is so obvious. Perhaps the use of the method of synthesis is also much harder in moving up the various levels of organization in biology because of the complexity of these systems. But both of these differences between biological phenomena and purely physical phenomena are only differences in degree, not in kind.

REDUCTION AND QUANTUM PHYSICS

Most of the preceding objections to reduction have been voiced by biologists and stem from their knowledge of biology and biological theories. But numerous physicists have also raised objections to the notion of reduction. Because quantum physics deals with natural phenomena at their very smallest, these objections have tended to revolve around implications of quantum physics for biology. Physicists from Erwin Schrödinger, Niels Bohr, Max Delbrück, and Werner Heisenberg to Eugene Wigner, Michael Polanyi, and Walter Elsasser have thought that quantum physics might have some significant implications for biology. On the face of it, this supposition is extremely implausible, because the special peculiarities of subatomic phenomena have little in the way of consequences for any macroscopic objects. Strangely enough, these physicists have been more extreme in their opposition to reductionism than have biologists of a similar persuasion.

One of the earliest and most ill-advised assertions concerning the relation between physics and biology arose from classical thermodynamics. According to the second law of thermodynamics, the overall course of the universe is toward increased entropy; i.e., toward decreased order. The continued existence of highly organized living creatures seems to contradict this principle. However, the existence of temporary pockets of increased order is completely compatible with the second law of thermodynamics. Living organisms maintain their organization at the expense of energy from their environment.

The two principles from quantum physics that supposedly have important consequences for biology are Heisenberg's principle of uncertainty (or indeterminacy) and Bohr's principle of complementarity. For centuries, physicists were determinists in the sense that all scientific laws had to be universal in form. Any departure from strict universality had to be due to insufficient evidence or failure to pursue the investigation to the appropriate level of analysis. If deterministic laws could not be found for phenomena at one level of analysis, physicists assumed that they would eventually discover them at a lower level. Absolute accuracy was perhaps impossible, but nothing stood in the way of increasing accuracy in observation indefinitely. However, at the subatomic level, these two assumptions proved to be mistaken. According to Heisenberg, it is impossible to determine simultaneously the position and momentum of a subatomic particle. If one of these parameters is ascertained, only a probability estimate can be obtained for the other. The laws of quantum physics are deterministic only for large ensembles of subatomic particles. They are necessarily statistical for single particles.

The principle of complementarity is not on a par with the principle of indeterminacy. It is not so much a part of quantum mechanics as a comment about it. Certain characteristics of light seem to require a particulate interpretation of light; other characteristics seem to imply that it is wavelike. But neither a particulate nor a wave theory of light seems to be able to account for all the characteristics of light. According to Bohr's principle of complementarity, these two theories of light are not reducible to one another, but they complement each other. In order to understand the nature of light completely, both interpretations, incompatible though they may be as models for light, must be used.

Quantum physics has at least one legitimate, though extremely indirect, consequence for biology. To some extent it has loosened the hold on the minds of philosophers and scientists of the absolute primacy of deterministic laws. If the basic laws of quantum physics are necessarily statistical, then perhaps the prevalence of statistical laws in biology is not all that bad. It should be kept in mind, however, that the statistical nature of quantum physics has been shown to be an inherent feature of current formulations of quantum theory. To eliminate it, quantum theory would have to be completely reformulated, and so far such a formulation has successfully eluded physicists. In Chapter Two we set out some of the reasons for the statistical nature of most formulations of evolutionary theory. Except for Mayr's founder principle, they involve the kind of contingency compatible with evolutionary theory being formulated in the traditional deductive manner.

But physicists were not content with the preceding implication of quantum physics for biology. They pushed the analogy even further. Niels Bohr (in Blackburn, 1966) argued that the necessity of keeping the object of investigation alive in biology results in a situation not unlike uncertainty at the quantum level. "Thus, we should doubtless kill an animal if we tried to carry the investigation of its organs so far that we could describe the role played by single atoms in vital functions. In every experiment on living organisms, there must remain an uncertainty as regards the physical conditions to which they are subjected, and the idea suggests itself that the minimal freedom we must allow the organism in this respect is just large enough to permit it, so to say, to hide its ultimate secrets from us." Pushing the analogy even further, Delbrück (in Blackburn, 1966) suggested that perhaps certain features of the living cell might stand in a complementary relationship to the underlying atomic structures. Neither knowledge of the biological characteristics of a cell alone nor knowledge of the molecular characteristics of a cell alone is adequate for understanding the nature of life. Instead complementary knowledge of the two levels is necessary.

As entertaining as the preceding observations might be, they do not stand up to careful scrutiny. Biologists carry on their investigations both *in vitro* and *in vivo*. Even when they are experimenting on a living specimen, they can investigate its organization quite extensively without killing it. Of course, they cannot discover all they wish to know using a single specimen; instead they make use of hundreds of laboratory specimens readily available. The applicability of the principle of complementarity to biological phenomena is even more tenuous. It is not formally a part of quantum physics, it is seriously questioned by many physicists, and it refers to two different interpretations of light at the same level of analysis. In the biological analogy, the issue concerns the relation between knowledge of living organisms obtained by investigations at the cellular level or higher and knowledge obtained by biochemical investigations. In a sense, such knowledge can be said to be "complementary," but not in the sense used in quantum physics.

COMPLEXITY AND UNIQUENESS

The preceding were exceedingly casual observations made by physicists about possible analogies between physical theories, primarily quantum physics and biology. In a series of publications extending over two decades, Walter Elsasser[4] has pursued these analogies even further, producing what he considers to be a conclusive argument against the reduction of biology to quantum physics. He argues that the classes of entities in quantum mechanics are homogeneous; that is, all the elements in a class are strictly indistinguishable. For example, except for spatial and temporal location, all atoms in the same quantum state are identical. Hence, though finite, such classes can be treated as if they were infinite. For the purposes of scientific investigation, one individual of a homogeneous class will do as well as another. Elsasser asserts that no such classes can be found in biology. Classes of biological entities are radically inhomogeneous, unique. Elsasser goes so far as to distinguish physics from biology by means of the presence of homogeneous classes in the former but not in the latter. He views "physics as the science dealing essentially with homogeneous systems and classes, and biology as the science of inhomogeneous systems and classes." Thus, if he is correct, the biologist is placed in a dilemma. The amount of experimentation necessary to find out all he needs to know would surely kill the organism under investigation, but he cannot go on to supplement the limited information obtained from the study of one specimen by investigating other members of the same class because they will all be different.

[4] *Atom and Organism* (Princeton: Princeton University Press, 1966); see criticism by Stuart Kauffman, "What Can We Know about a Metazoan's Entire Control System?: on Elsasser's and Other Epistemological Problems in Cell Science," *Towards a Theoretical Biology,* Vol. IV (Chicago: Aldine Publishing Co., 1972).

If Elsasser's argument were cogent, it would be impossible to reduce biological theories to those of physics and chemistry because there could be no biological theories. By the same token, there would be no theories in physics save quantum theory. Only at the subatomic level can scientists distinguish significant classes whose members are strictly indistinguishable. Even at the level of atoms, problem cases begin to arise. No sooner does Elsasser claim that one "of the most remarkable properties of atoms and of molecules of a given species is that they are exactly alike," than he is forced to recognize the existence of isotopes, atoms that contain additional neutrons. (For example, heavy water is water made up of a large percentage of water molecules that contain extra neutrons.) Elsasser's claim for physics is further invalidated by the existence of isomers—molecules that have identically the same structure but are mirror images of each other. At the level of macroscopic objects, Elsasser's thesis about physics completely dissolves. As dreary as they may seem to the casual observer, no two Volkswagons are indistinguishable in the sense that two protons are. If homogeneous classes are necessary for science, then quantum physics is the only science.

Surely Elsasser has overstated his case, and in one place he seems to realize it, admitting that "by the very nature of a class, its members have certain properties in common (members of a class of organisms as a rule have very many properties in common), and hence a partial prediction based upon the common properties of the members of the class can always be achieved." Even in the above admission, Elsasser makes several unwarranted assumptions. For instance, he assumes that the names of classes of organisms can be defined in terms of sets of characters which are severally necessary and jointly sufficient for membership. As we have argued earlier, this assumption may be appropriate for certain classes of organisms taken as atemporal time slices, but not for all, and if a temporal dimension is admitted, then only those species that have evolved saltatively can be defined in the way required by Elsasser.

In addition, Elsasser assumes that the elements of any significant biological class will be organisms. Organisms are very complex systems. An inverse relation exists between complexity and identity. The more complex a system is, the more difficult it is to find another system identical to it in every respect. The closest thing that can be found in biology to Elsasser's homogeneous classes are clones, individuals derived from a single complement of chromosomes. Even in this most extreme case, members of a clone vary somewhat, both genetically because of imperfections in the hereditary mechanisms and phenotypically because of the variable influence of the environment.

Contrary to Elsasser's beliefs, science is possible even without strictly homogeneous classes. Physicists do not study the heavenly bodies in all their complexity. Instead they try to find certain properties these bodies have in common that exhibit significant relations with one another. For example,

Newton's laws deal only with the mass and velocity of material bodies, not with their color, shape, or chemical composition. Similarly, it is a mistake to think that biology must deal with each organism in all its complexity. Like all scientists, biologists try to discern significant properties of living organisms that can be integrated into scientific laws and theories. Finding the right properties and combining them in the right way is the chief task of the theoretical scientist. For example, in evolution, what is significant—the actual number of organisms in a species, relative densities, the number of different species represented, or what? As was observed in Chapter Two, one of the crucial questions in biology is whether the degree of precision necessary in the specification of the genetic makeup of populations is so great as to preclude the formulation of any evolutionary laws about them.

THE ROLE OF BOUNDARY CONDITIONS IN SCIENCE

Polanyi has offered a proof that "both machines and living mechanisms are irreducible to the laws of physics and chemistry."[5] A machine "as a whole works under the control of two distinct principles. The higher one is the principle of the machine's design, and this harnesses the lower one, which consists in the physical-chemical processes on which the machine relies." Similarly, an "organism is shown to be, like a machine, a system which works according to two different principles: its structure serves as a boundary condition harnessing the physical-chemical processes by which its organs perform their functions." When Polanyi says that higher-level principles harness lower-level principles and/or phenomena, he seems to mean that the existence of structure at one level imposes boundary conditions for the laws operating at lower levels. He then argues further that because a "boundary condition is always extraneous to the process which it delimits," any structured system, specifically the structure of machines and the morphology of living things, "transcends the laws of physics and chemistry."

Because of the extreme looseness with which Polanyi uses his already misleading terminology, he seems to be reifying his principles, attributing causal efficacy to such things as designs, vocabularies, grammars, and information. Sometimes his principles harness lower-level phenomena, sometimes lower-level laws. For example, in one place he says "we harness the laws of nature" when we construct a machine, in another that the "structure of the machine" harnesses the "forces of nature." In another, he writes that the "principles of the machine's design" do the harnessing. The explanatory example he gives of boundary conditions is the slope of the inclined planes in Galileo's experiments. He observes correctly that the choice of the slope

[5] "Life's Irreducible Structure," *Science* 160 (1968), 1308–12; see also Ronald Giere's reply in *Science,* 162 (1968), 410, and Simon, 1971.

does not follow from the laws of mechanics. Similarly, the positions and momenta of the heavenly bodies at a particular instant are the boundary conditions for the solar system in celestial mechanics. Unfortunately, too many of the things Polanyi mentions cannot cogently be called boundary conditions, given his explanatory example. If the slope of an inclined plane is a boundary condition, then it is unlikely that he can extend this analogy to include the mind. Although he begins with machines and progresses to the morphology of living organisms, his chief interest seems to be the status of mind. Like all the other things already mentioned, "mind harnesses neurophysiological mechanisms and is not determined by them." It is difficult to see how the mind can serve as a boundary condition for neurophysiological mechanisms.

Polanyi recognizes that the analogy between machines and organisms is weakened somewhat by differences in their origin. Human beings construct machines and are responsible for their structure. The structure of living organisms arises through the evolutionary process. As might be expected, he finds the currently popular version of evolutionary theory sorely deficient. Structure cannot arise through chance variation and natural selection. "No highly significant order can ever be said to be solely due to an accidental collocation of atoms." The synthetic theory of evolution must be supplemented by the "operation of an ordering principle." This *"ordering principle* which *originated* life is the *potentiality* of a stable open system."[6]

Evolutionary theory is easily caricatured. One of the commonest parodies is that, according to evolutionary theory, life arose through the accidental collocation of atoms, as if atoms were so many marbles bouncing around in empty space. But atoms possess bonding properties. Certain atoms are more likely to combine in certain ways than are others. Such knowledge is commonplace. Any arguments about the likelihood of the origin of life which ignore these facts count neither against the synthetic theory of evolution nor against the mechanistic world view. The only candidate for Polanyi's ordering principle that originated life is the bonding properties of the chemical elements.

In spite of the numerous infelicities in Polanyi's argument, the conclusion he seems to be urging, once it is clearly formulated, is important. Once a living system has evolved to a certain level of complexity so that the genotype-phenotype distinction can be made, future development is constrained by the organization of the genotype, the organization of the phenotype, and the complex relation between the two. The occurrence of mutations may be independent of such considerations, but the establishment of a mutation is not. However, Polanyi is not saying anything new. Biologists from Darwin to Mayr have emphasized the need for maintaining sufficient organization

[6] Polanyi, *Personal Knowledge* (Chicago: University of Chicago Press, 1958), p. 35.

in both the genome and the phenome to permit reproduction. Even though Fisher and Haldane were forced to consider a single locus at a time in their early investigations of the mechanisms of evolution, they were well aware that "in a realistic model of evolution we cannot deal with each gene locus as an isolated phenomenon. It is a gross exaggeration to claim that every gene affects all characters. Yet this statement is probably closer to the truth than the belief of early Mendelians."[7]

Finally, Polanyi makes the important observation that a scientist's interest is not always limited to scientific laws. Sometimes he is more interested in the boundary conditions themselves. "When a saucepan bounds a soup that we are cooking, we are interested in the soup; and, likewise, when we observe a reaction in a test tube, we are studying the reaction, not the test tube. The reverse is true for a game of chess. The strategy of the player imposes boundaries on the several moves, which follow the laws of chess, but our interest lies in boundaries—that is, in the strategy, not in the several moves as exemplifications of the laws." Evolutionists are frequently more interested in the boundary conditions in the evolutionary process than in the laws. As we observed in Chapter Three, the laws that biologists use to aid in the reconstruction of phylogenetic development are so weak that the data carry most of the explanatory weight. In the case of relatively closed systems like the solar system, all we need to know in order to predict the future states of the system are the laws of celestial mechanics and the positions and momenta of the various solar bodies at any one time. Hence, our interest tends to center on the laws. In the case of partially open systems, such as organisms and species, we have to know which mutations occur, when and in what order, the effects of these mutations on the phenotype, and the successive changes in the environment. Given current biological, geological, and meteorological theories, a specification of these variables at any one time will not do. We need to obtain additional data frequently, so frequently that the specification of boundary conditions in evolutionary studies takes on the character of an historical narrative (see Chapter Three). Those who might feel that such a heavy reliance on boundary conditions puts evolutionary explanations in a bad light should recall that a major school of physicists traces one of the most fundamental notions in physics, the direction of time, back to the boundary conditions that gave rise to the present expansion of the universe.

[7] Ernst Mayr, in *Mathematical Challenges to the Neo-Darwinian Interpretation of Evolution* (Philadelphia: The Wistar Institute Press, 1967), p. 53.

Conclusion

By now I need not tell the reader that the problems in the philosophy of biology are both difficult and very closely interconnected. Biologists have emphasized the complexity and consequent uniqueness of biological systems, their stratified, teleological organization, and the central role of historical considerations in biology. Before a system can behave teleologically, it must attain a certain degree of complexity. Except for certain artifacts, systems come to possess this requisite degree of complexity only through selection processes. In the preceding pages, I have tried to explain what biologists mean by such claims. I cannot pretend that the issues discussed are easy to comprehend or that I have explained them as clearly as might be possible. My hope is that in this book I have brought together the most significant philosophical problems in biology and shown their interrelations with sufficient clarity to permit others to provide better solutions than I have been able to suggest in this short book.

FOR FURTHER READING

ANTHOLOGIES

APPLEMAN, PHILIP, *Darwin, A Norton Critical Edition.* New York: W. W. Norton & Co., 1970.

BLACKBURN, ROBERT T., ed., *Interrelations: The Biological and Physical Sciences.* Chicago: Scott, Foresman and Co., 1966.

CANFIELD, JOHN V., ed., *Purpose in Nature.* Englewood Cliffs, N. J.: Prentice-Hall, Inc., 1966.

CARLSON, ELOF AXEL, ed., *Modern Biology: Its Conceptual Foundations.* New York: Braziller, 1967.

GABRIEL, MORDECAI L., and SEYMOUR FOGEL, eds., *Great Experiments in Biology.* Englewood Cliffs, N.J.: Prentice-Hall, Inc., 1955.

GLASS, BENTLEY, OWSEI TEMKIN, and WILLIAM L. STRAUS, JR., eds., *Forerunners of Darwin: 1745–1859.* Baltimore: The Johns Hopkins Press, 1959.

HINTIKKA, JAAKKO, ed., *Synthese* (vol. 20, no. 2 devoted to philosophy of biology), August 1969.

HULL, DAVID L., ed., *Darwin and His Critics: The Reception of Darwin's Theory of Evolution by the Scientific Community.* Cambridge: Harvard University Press, 1973.

MENDELSOHN, EVERETT, ed., *Journal of the History of Biology* (vol. 2, no. 1, devoted to history and philosophy of biology), Spring 1969.

MUNSON, RONALD, ed., *Man and Nature.* New York: Delta Books, 1971.

PETERS, JAMES A., ed., *Classic Papers in Genetics.* Englewood Cliffs, N. J.: Prentice-Hall, Inc., 1959.

BOOKS BY INDIVIDUAL AUTHORS

BECK, WILLIAM S., *Modern Science and the Nature of Life.* Garden City, N.J.: Anchor Books, 1961.

BECKNER, MORTON, *The Biological Way of Thought.* Berkeley: University of California Press, 1959.

BOSSERT, W. H., and E. O. WILSON, *A Primer of Population Biology.* Stamford, Conn.: Sinauer Assoc., 1972.

CARLSON, ELOF AXEL, *The Gene: A Critical History.* Philadelphia: W. B. Saunders Company, 1966.

DOBZHANSKY, THEODOSIUS, *Genetics of the Evolutionary Process.* New York: Columbia University Press, 1970.

ELLEGÅRD, ALVAR, *Darwin and the General Reader: The Reception of Darwin's Theory of Evolution in the British Periodical Press, 1859–1872.* Gothenburg: Göteborgs Universitets Årskrift, 1958.

GHISELIN, MICHAEL T., *The Triumph of the Darwinian Method.* Berkeley: University of California Press, 1969.

GOUDGE, T. A., *The Ascent of Life.* Toronto: University of Toronto Press, 1961.

HEMPEL, CARL G., *Aspects of Scientific Explanation.* New York: The Free Press, 1965.

LEVINS, RICHARD, *Evolution in Changing Environments.* Princeton: Princeton University Press, 1968.

LOVEJOY, ARTHUR O., *The Great Chain of Being.* New York: a Harper Torchbook, 1936.

MACARTHUR, ROBERT H., *Geographical Ecology.* New York: Harper and Row, 1972.

MAYR, ERNST, *Animal Species and Evolution.* Cambridge: Harvard University Press, 1963.

OLBY, ROBERT, *Origins of Mendelism.* New York: Schocken Books, 1966.

PROVINE, WILLIAM B., *The Origins of Theoretical Population Genetics.* Chicago: The University of Chicago Press, 1971.

RUSE, MICHAEL, *Philosophy of Biology.* London: Hutchinson's University Library, 1973.

SIMON, MICHAEL A., *The Matter of Life.* New Haven: Yale University Press, 1971.

SIMPSON, GEORGE GAYLORD, *This View of Life.* New York: Harcourt, Brace & World, Inc., 1964.

SMART, J. J. C., *Philosophy and Scientific Realism.* London: Routledge & Kegan Paul, 1963.

———, *Between Science and Philosophy.* New York: Random House, 1968.

VORZIMMER, PETER, *Charles Darwin. The Years of Controversy. The Origin of Species and its Critics.* Philadelphia: Temple University Press, 1970.

WATSON, JAMES D., *Molecular Biology of the Gene.* New York: W. A. Benjamin, Inc., 1970.

WILLIAMS, G. C., *Adaptation and Natural Selection.* Princeton: Princeton University Press, 1966.

INDEX